Martin Nirschl

Label-free Biosensors: Thin-film Bulk Acoustic Resonators

Martin Nirschl

Label-free Biosensors: Thin-film Bulk Acoustic Resonators

Theory and Application of FBARs for Biomolecular Interaction Analysis

Südwestdeutscher Verlag für Hochschulschriften

Impressum/Imprint (nur für Deutschland/only for Germany)
Bibliografische Information der Deutschen Nationalbibliothek: Die Deutsche Nationalbibliothek verzeichnet diese Publikation in der Deutschen Nationalbibliografie; detaillierte bibliografische Daten sind im Internet über http://dnb.d-nb.de abrufbar.
Alle in diesem Buch genannten Marken und Produktnamen unterliegen warenzeichen-, marken- oder patentrechtlichem Schutz bzw. sind Warenzeichen oder eingetragene Warenzeichen der jeweiligen Inhaber. Die Wiedergabe von Marken, Produktnamen, Gebrauchsnamen, Handelsnamen, Warenbezeichnungen u.s.w. in diesem Werk berechtigt auch ohne besondere Kennzeichnung nicht zu der Annahme, dass solche Namen im Sinne der Warenzeichen- und Markenschutzgesetzgebung als frei zu betrachten wären und daher von jedermann benutzt werden dürften.

Coverbild: www.ingimage.com

Verlag: Südwestdeutscher Verlag für Hochschulschriften GmbH & Co. KG
Heinrich-Böcking-Str. 6-8, 66121 Saarbrücken, Deutschland
Telefon +49 681 37 20 271-1, Telefax +49 681 37 20 271-0
Email: info@svh-verlag.de

Approved by: Zürich, ETH, Diss., 2011

Herstellung in Deutschland:
Schaltungsdienst Lange o.H.G., Berlin
Books on Demand GmbH, Norderstedt
Reha GmbH, Saarbrücken
Amazon Distribution GmbH, Leipzig
ISBN: 978-3-8381-3165-8

Imprint (only for USA, GB)
Bibliographic information published by the Deutsche Nationalbibliothek: The Deutsche Nationalbibliothek lists this publication in the Deutsche Nationalbibliografie; detailed bibliographic data are available in the Internet at http://dnb.d-nb.de.
Any brand names and product names mentioned in this book are subject to trademark, brand or patent protection and are trademarks or registered trademarks of their respective holders. The use of brand names, product names, common names, trade names, product descriptions etc. even without a particular marking in this works is in no way to be construed to mean that such names may be regarded as unrestricted in respect of trademark and brand protection legislation and could thus be used by anyone.

Cover image: www.ingimage.com

Publisher: Südwestdeutscher Verlag für Hochschulschriften GmbH & Co. KG
Heinrich-Böcking-Str. 6-8, 66121 Saarbrücken, Germany
Phone +49 681 37 20 271-1, Fax +49 681 37 20 271-0
Email: info@svh-verlag.de

Printed in the U.S.A.
Printed in the U.K. by (see last page)
ISBN: 978-3-8381-3165-8

Copyright © 2012 by the author and Südwestdeutscher Verlag für Hochschulschriften GmbH & Co. KG and licensors
All rights reserved. Saarbrücken 2012

Dedication

I would like to dedicate this book to Lele, Marta, Paola, Mama and Papa.

Acknowledgements

During my PhD I had the pleasure of meeting numerous people who helped me with my thesis. Working together with them and the variety of their personalities made it much easier to overcome the unavoidable dark, frustrating and boring phases of the PhD.
I would like to thank the following individuals:

From Siemens AG:
I would like to thank my supervisor **Matthias Schreiter** for giving me the opportunity of conducting my research as part of his team. His continuous support during the years, the integration into running projects combined with the possibility to bring in my own ideas and the freedom to conduct own measurements were most important for this thesis.
I am especially thankful to the "clean-room team" **Dana Pitzer, Thomas Huber and Robert Primig** together with all the students with whom I worked, especially **Régulo Miguel Ramírez Wong, Özlem Karaca, Brigitte Lechner, Luis Resenberg** and **Agnes Agnieszka**: Their interest, commitment and work were essential for many of the results described in this thesis. I would also like to thank **Willi Metzger** for the numerous on- and off topic chats and his help with simulations and organisational work. A special thanks goes to **Daniel Sickert** for his numerous troubleshooting of the bio- and micro world, our measurements together, and for being one of the last people who believes in the future of CNTs.
Special thanks goes to **Florian Reuter** (Siemens Technology Accelerator GmbH) who provided valuable support, help and precious motivation in the final stage of my work and, furthermore, for the great journeys we made together.

From ETH Zürich:
I would like to thank my supervisor **Janos Vörös** for accepting me as his student in his group. His scientific input and excitement about my topic was necessary for making large parts of this work possible. Furthermore I would like to thank him for his patience while correcting my papers and for the injection of motivation coming from every single discussion we had. I would also like to thank **the whole LBB group at the ETH**, especially

Marta Bally, **Dorothee Grieshaber** and **Raphael Zahn** for helping me with the vesicles and the polyelectrolytes.

From VTT Finland in Tampere I would like to thank **Inger Vikholm-Lundin** and **Sanna Auer** for our great collaboration, the help with and providing the DNA surface chemistry and our measurements together.

From VTT Finland in Espoo I would like to thank **Kari Tukkiemi** and **Arto Rantala** for the great collaboration, the help with any problems related with the FBAR read-out and the whole group for the fantastic hands-on explanation of Finish sauna rules.

From Biosensor Applications AB I would like to **Ann-Charlotte Hellgren** for the measurements we made together and the whole Biosensor Applications AB team for the invitation to their place.

From TU Dresden I would like to thank **Anja Blüher** for her support and patience with the measurements we made together and **Oliver Jost** for providing the carbon nanotubes and the SEM picture.

From Cranfield University I would like to thank **Dimitris Kyprianou** for providing his surface chemistry.

I would like to thank **Johannes Ottl** from Novartis and **Manfred Auer** from the University of Edinburgh for their interest in the FBAR technology and support with our measurements.

From the Farfield Group I would like to thank **Marcus Swann** and **Gerry Ronan** for sharing measurement data and for the invitation to their place.

From Chalmers University I would like to thank **Fredrik Höök** for our helpful discussion and for being co-referee in my thesis defense.

I would like to thank **David Evans** (SESMOS Ltd., Edinburgh, Scotland) for linguistic editing and proofreading my thesis.

Abstract

While FBARs have been used for decades in mobile communications and hundreds of millions have been sold for use in mobile phones or GPS devices, their usage as a sensor is rather novel. Because the FBAR's attributes such as its resonant frequency change if mass is added on top, the FBAR can be used for any application where an absorbate on top needs to be detected or quantified. FBAR sensors can work in air or in liquid. FBAR sensors in air could be used as gas sensors (e.g. humidity or CO_2 sensors), detectors for explosives or drugs or as so-called electronic noses, which detect volatile organic compounds (VOC) such as odours.

FBARs might be used for applications in liquid such as the detection of bacteria or the monitoring of water quality or liquids in industrial processes. Furthermore, the FBAR can be used to detect substances such as proteins or other types of disease marker in bodily fluids (e.g. blood or saliva). This might be useful in either diagnostics or for point-of-care devices providing they fulfil certain criteria such as sufficient sensitivity. An example of a point-of-care device would be a handheld device that a patient could use to determine information about his or her health status from a drop of blood. The device would measure the concentration of certain substances in the blood, for example cancer markers. A further application might be the area of drug discovery and drug development. In this area the FBAR could be used to monitor the interaction between molecules such as a drug candidate and a disease target.

This thesis concentrates on monitoring interactions between macromolecules such as proteins or DNA using the FBAR, as they currently look like the most promising types of application. The main focus was the development of the technology towards applications in drug discovery/development, a field in which the FBAR technology appears extremely promising.

To this end, two main goals were established: Firstly, the understanding of the design rules for the FBAR as a mass sensor, and their utilization in improving the design, and secondly the use of the FBAR in performing measurements closer to potential applications. These two goals are strongly coupled; the design considerations inform how tests for potential applications should be performed and practical experience from the application tests can be used to improve the design process.

To learn about the FBAR behaviour a variety of solid materials was deposited on top. The thicknesses of the material were varied from being thin compared to the resonator to thicknesses that are in the range of the thickness of the resonator (i.e. from few nanometres up to some hundreds of nanometres) and thus overload the FBAR. Conclusions about the acoustic properties of the adsorbate were then drawn from the resonator response. E.g. in a test measurement at a resonant frequency of around 800 MHz it was found that the acoustic properties of carbon nanotubes (CNTs) are interesting as a coating material in the FBAR due to their low acoustic impedance. Using materials with low acoustic impedance generally increases the mass sensitivity of acoustic devices.

The first test of applications consisted of protein detection using an appropriate antibody as a surface functionalisation. Similarly, specific sequences of short DNA segments were also detected: For this, the FBAR was functionalised with strands complementary to the target DNA molecules. With two different functionalisations available, two different DNA sequences could be detected selectively from buffer and diluted human blood serum.

As a first demonstration of multiplexed measurement, two resonators functionalised with different antibodies were simultaneously read-out. When the corresponding antigens were added one after the other, only the resonator with the complementary functionalisation showed a binding signal.

In another measurement, the adsorption of S Layer proteins to the gold surface of the sensor was studied. A concentration-dependent adsorption behaviour was found and in the long term measurement (> 2h) a recrystallisation process was detected.

By comparing the FBAR measurement results of the absorbtion of both a lipid bilayer and a polyelectrolyte multilayer with the results obtained by quartz crystal microbalance, conclusions about the influence of a higher operating frequency could be drawn.

For thick films it was found that, due to the approximately two order of magnitude higher resonant frequency, the FBAR is significantly more sensitive to viscoelastic properties of the adsorbate than to changes in layer thickness.

The FBAR reached a new level when it was combined with a CMOS read-out system. This allowed the simultaneous read-out of 64 resonators. In a first measurement all resonators were functionalised individually using a nanospotter with different single stranded DNA strands and the hybridisation of the complementary sequences was followed on all pixels in real-time.

The last part of the thesis describes an initial attempt to demonstrate that it is possible to monitor conformation changes with the FBAR. The conformational changes of a model protein, calmodulin, upon subsequent binding of calcium and a peptide were investigated. From the FBAR response both mass adsorption and a conformational change of the calmodulin were visible.

The simulations and measurements in this thesis improved the understanding of the device and the close-to-application experiments helped to evaluate the performance of the FBAR as a biosensor.

Table of Contents

1 Introduction to Biomolecular Interaction Analysis 15
1.1 Transducer Principles for Biomolecular Interaction Analysis 15
    1.1.1 Acoustic Sensors 17
    1.1.2 Optical Sensors 21
    1.1.3 Isothermal Titration Calorimetry (ITC) 29
    1.1.4 Electrochemical Sensors 30
    1.1.5 Nanostructure Biosensors 32
1.2 Label-free Transducer Principles to Investigate Conformational Changes 36
    1.2.1 QCM/QCM-D 36
    1.2.2 Dual Polarisation Interferometry (DPI) 37
    1.2.3 Backscattering Interferometry (BSI) 39
    1.2.4 Isothermal Calorimetry (ITC) and Differential Scanning Calorimetry (DSC) 41
1.3 Conclusion and Outlook 43

2 Thesis Scope 45

3 Materials and Methods 48
3.1 Chemicals, Proteins, Polymers & DNA 48
3.2 The FBAR and the Measurement Set-up 50
3.3 QCM 53
3.4 SPR 54
3.5 Simulations of the Electrical Impedance Spectrum, Frequency Shift and Mass Sensitivity for QCM and FBAR 54
3.6 Functionalisation for DNA detection 55

4 The influence of the acoustic properties of the sensor materials on the FBAR performance 57
4.1 Experimental Section 58
    4.1.1 FBAR 58
    4.1.2 Thin-film Deposition 59
    4.1.3 Carbon Nanotube Films 59

4.2 Simulation of the Frequency Response for Thin-film Adsorption of Materials with Different Acoustic Velocities and Mass Densities 60
4.3 Deposition of Platinum, Aluminium Oxide, and Tungsten Thin-films 65
4.4 Deposition of Carbon Nanotube Films 67
4.5 Enhancement of FBAR Mass Sensitivity using Materials with Low Acoustic Impedance such as CNTs 69
4.6 Summary and Outlook 70

5 Measurement of DNA and Protein Adsorption on Passive FBAR 72
5.1 Experimental Section 72
    5.1.1 S Layer Proteins 72
5.2 FBAR Mass Sensitivity Comparison with SPR 73
5.3 Measurement of Protein-Protein Interaction using a Dual Measurement Probe 74
5.4 DNA Detection 76
5.5 Adsorption an Recrystallisation of S-Layer Proteins on Gold 81
5.6 Conclusion and Outlook 85

6 Lipid Bilayer and PEM 87
6.1 Experimental Section 88
    6.1.1 Lipids and Vesicles 88
    6.1.2 Polyelectrolyte Multilayers (PFM) 88
6.2 Vesicle Adsorption and Bilayer Formation 89
6.3 Polyelectrolyte Multilayer (PEM) 91
6.4 Discussion 93
6.5 Conclusions 97

7 CMOS-integrated FBAR Array for Specific and Selective Multiplexed Detection of DNA in Buffer and Diluted Serum 99
7.1 Experimental Section 100
    7.1.1 CMOS-integrated FBAR 100
    7.1.2 BSA Measurements 101
7.2 Mass Sensitivity Comparison obtained with FBAR, SPR and QCM 102
7.3 Multiplexed DNA Measurement 104
7.4 Multiplexed Measurement of PCR Amplified Products in Buffer and Serum 106
7.5 Combination of Compartments and Piezo Dispenser for Label-free Biosensing 109
7.6 Conclusion 114

8   Conformational change of Calmodulin 115
8.1   Introduction 115
8.2   Experimental Section 116
    8.2.1   FBAR Read-out 116
    8.2.2   Reagents and Materials 116
8.3   Results and Discussions 116
    8.3.1   Immobilisation of Neutravidin and Biotinylated Calmodulin 116
    8.3.2   Calcium Induced Conformational Changes of Calmodulin 117
    8.3.3   CaMKII Peptide Binding to the Ca^{2+}/Calmodulin Complex 123
    8.3.4   Conductance versus Frequency Shift Plots 125
8.4   Conclusion 127
9   Conclusions and Outlook 129

Abbreviations

Here is a list of the most commonly mentioned abbreviations in this thesis:

ApoCaM	Apocalmodulin (i.e. Calcium-free Calmodulin)
BAW	Bulk Acoustic Wave
BIA	Biomolecular Interaction Analysis
BLI	Bio-Layer Interferometry
BSA	Bovine Serum Albumin
BSI	Backscattering interferometry
CCD	Charge-Coupled Device
CMOS	Complementary Metal–Oxide–Semiconductor
CNT	Carbon Nanotube
DMSO	Dimethyl Sulfoxide
DMT	Dimethyl terephthalate
DNA	Deoxyribonucleic Acid
DOC	Deoxycholic Acid
DOPC	1,2-Di-dleoyl-sn-glycero-3-Phosphocholine
DPI	Dual Polarization Interferometry
DSC	Differential Scanning Calorimetry
EDTA	Ethylenediaminetetraacetic Acid
EIS	Electrochemical Impedance Spectroscopy
ELM	Ellipsometry
FBAR	(Thin-) Film Bulk Acoustic Resonator
FET	Field Effect Transistors
FPGA	Field-Programmable Gate Array
GPS	Global Positioning System

HEPES	4-(2-hydroxyethyl)-1-piperazineethanesulfonic acid
HPLC	High-performance liquid chromatography
HTS	High-throughput screening
ITC	Isothermal Calorimetry
ITD	Interdigital Transducers
ITO	Indium Doped Tin Oxide
LASER	Light Amplification by Stimulated Emission of Radiation
LBL	Layer-by-Layer Deposition
LFE	Lateral Field Excitation
LOD	Limit of Detection
LSPR	Localized Surface Plasmon Resonance
MCH	6-mercapto-1-hexanol
NMR	Nuclear Magnetic Resonance
OWLS	Optical Waveguide Lightmode Spectroscopy
PAH	Poly(allylamine hydrochloride)
PBS	Phosphate Buffered Saline
PCB	Printed Circuit Board
PCR	Polymerase Chain Reaction
PDMS	Polydimethylsiloxane
PEI	Polyethyleneimine
PEM	Polyelectrolyte Multilayers
PGA	Poly(l-glutamic acid)
PLL	Poly(L-lysine)
QCM	Quartz Crystal Microbalance
QCM-D	Quartz Crystal Microbalance with Dissipation Monitoring

RNA	Ribonucleic acid
SAW	Surface Acoustic Wave
SDS	Sodium Dodecyl Sulphate
SEM	Scanning Electron Microscope
SHW	Shear Horizontal Waves
SIMS	Secondary Ion Mass Spectrometry
SMR	Surface Mounted Resonator
SPR	Surface Plasmon Resonance
STW	Surface Transverse Waves
TRIS	Tris(hydroxymethyl)aminomethane
TSM	Transverse Shear Mode
VCO	Voltage-Controlled Oscillator
VOC	Volatile Organic Compound

1 Introduction to Biomolecular Interaction Analysis[1]

Label-free biomolecular interaction analysis is an important technique for the study of chemical binding between biomolcules such as proteins or between a protein and a small molecule in real-time. The parameters obtained with this technique, such as the affinity, are important for drug development. While surface plasmon resonance (SPR) instruments are the most widely used, new types of sensor are emerging. These developments are generally driven by the need for higher throughput, lower sample consumption or to provide complimentary information to the SPR data. This introduction aims to give an overview of a wide range of sensor transducers, the working principles and the specific limitations of each technology, e.g. concerning the set-up, sensitivity and limit of detection (LoD), sensor size or required sample volume. Starting from optical technologies like SPR and waveguide based sensors, acoustic sensors like the QCM and the FBAR, on which this thesis is focused, calorimetric and electrochemical sensors are covered. Technologies long established in the market are presented together with those newly commercially available and with technologies in the early stages of development.

Finally, the commercially available instruments are summarized together with their sensitivity and the number of sensors usable in parallel and an outlook for potential future developments is given.

1.1 Transducer Principles for Biomolecular Interaction Analysis

Biomolecular interaction analysis (BIA) is an important method for drug discovery and drug development [1]. Label-free sensors have the advantage that the adsorbed molecules do not require any chemical treatment such as the addition of a radioactive, fluorescent or other type of marker[2]. This saves resources and is a significant advantage for the study of

---

[1] Parts of this chapter are included in Nirschl, M.; Reuter, F.; Vörös, J., Review of Transducer Principles for Label-Free Biomolecular Interaction Analysis. *Biosensors* **2011**, *1*, 70-92.
[2] An exeption is a sandwich assay, where substances are not detected directly but via a molecule that specifically binds to the substance to be detected and is easier to be measured.

the interaction between molecules due to the potential for the presence of a label altering the interaction process.

Several parameters are important when selecting a transducer to use for BIA: The most obvious is the **limit of detection (LOD)**, which can be defined (i) as the smallest detectable concentration of a certain substance or (ii) the lowest detectable molecular mass of a certain concentration of molecules, (iii) the lowest detectable affinity of a chemical reaction or (iv) for surface-based sensors the lowest detectable surface mass density. As only the transducer principles should be compared here, the smallest detectable surface mass will be focused on because this measure is only dependent on the transducer. Other parameters like the smallest detectable concentration may depend strongly on factors independent of the transducer like the surface chemistry used or the fluidic system. However, parameters other than the sensitivity are also equally important: The required sample volume is crucial if many substances or many different concentrations are measured as in the case of high-throughput screening (HTS), if the sample is available in very limited volumes (e.g. human drug targets), or the transducer is integrated into other processes delivering small sample amounts like on-bead screening [2].

The number of parallel usable sensors is important if a high throughput in a short time is desired. The more multiplexed a sensor is, the more parallel measurements can be performed without significantly increasing the equipment size and cost. A wide range of transducer principles have been developed and used for BIA in the last few decades. This section aims to give an overview of the state-of-the-art of different transducers used for label-free BIA. The most important parameters are summarised in Table 1.

While this paper aims to give an overview of label-free transducers, there are detailed reviews available for acoustic [3, 4], optical [5-7], electrochemical [8] and nanostructure-based [9, 10] transducers.

Also of interest to the reader, might be reviews covering topics related to label-free biosensors with a special emphasis on highly multiplexed technologies [11, 12], microdispensing of liquids for biosensor arrays [13] and label-free cell-based assays in drug discovery [14].

1.1.1 Acoustic Sensors

Acoustic sensors comprise one or more vibrating elements that create acoustic waves. These waves can propagate on the surface, i.e. surface acoustic wave (SAW) or in the bulk of the resonator, i.e. bulk acoustic wave (BAW). The properties (e.g. amplitude or frequency) of these waves change when molecules adsorb and physically or chemically bind to the sensor surface. This change is detected and information about the amount of adsorbed molecules can be derived.

This overview of acoustic sensors is limited to acoustic sensors vibrating parallel to the sensor surface, as resonators vibrating vertically to the sensor surface (e.g. in the longitudinal mode) have a high loss of energy into the liquid and are limited in sensitivity and thus of limited use in monitoring adsorbates of biomolecules in real-time. An overview of all acoustic microsensors including cantilever-based sensors or micromachined ultrasonic transducers (CMUTs) can be found in [15].

Quartz Crystal Microbalance (QCM) and Quartz Crystal Microbalance with Dissipation Monitoring (QCM-D)

The QCM is a bulk acoustic wave (BAW) device, which consists of a piezoelectric quartz crystal, which resonates if electrically excited using two electrodes (Figure 1). Sauerbrey found that the resonant frequency decreased linearly if additional mass is attached to the sensor [16]. However, this is only true if the attached mass is rigid and small compared to the sensor mass. If the attached mass is not rigid, the viscoelastic properties have to be taken into account. This is usually the case for operation in liquids [17] and for the adsorption of soft materials. Using a model where the adsorbed soft material is represented by a viscous and an elastic element connected in parallel (i.e. a Kelvin-Voigt material) under a Newtonian liquid it is also possible to describe the frequency response in a liquid environment [18]. The frequency shift Δf which is influenced by the attached mass, the liquid environment around the sensor and the viscoelastic properties of the adsorbate is hereby given by:

$$\Delta f \approx -\frac{\eta_2}{2\pi n_q \delta_2(\omega)} - \frac{h_1 \rho_1 \omega}{2\pi n_q}\left[1 - \frac{2}{\rho_1}\left(\frac{\eta_2}{\delta_2(\omega)}\right)^2 \frac{G'(\omega)}{G'(\omega)^2 + G''(\omega)^2}\right] \tag{1}$$

with $\delta_2 = \sqrt{-2\eta_2/\rho_2\omega}$ where η is the viscosity, m the mass, ρ the density, ω the angular frequency, h the thickness of the adsorbate, G' the storage and G'' the loss modulus of the adsorbate. The subscript '1' corresponds to the adsorbed layer, the subscript 'q' to the quartz and the subscript '2' to the bulk liquid [19].

The QCM has a LOD lower than 1 ng/cm² and can also be used for adsorbates several hundred nanometres thick depending on their acoustic properties. Due to this high dynamic range the QCM is used in a broad range of applications, from small molecules up to cells [20].

More recently attention was not only drawn to measuring the adsorbed mass but also to investigate the viscoelastic properties of the adsorbate. This can be done by not only reading out the resonant frequency, but also the motional resistance [21], the conductance [22] or the energy dissipation [23]. The latter system is named quartz crystal microbalance with dissipation monitoring (QCM-D). With this technique, novel types of investigations like on the changes of viscoelastic properties of polymers [24], vesicle adsorption and lipid bilayer formation [25], cross-linking of protein layers [26] and folding or unfolding of proteins have been performed.

In most commercially available QCM systems a sample volume of few tens of μl is needed per flow cell, motivating the search for a smaller BAW device with smaller sensor area.

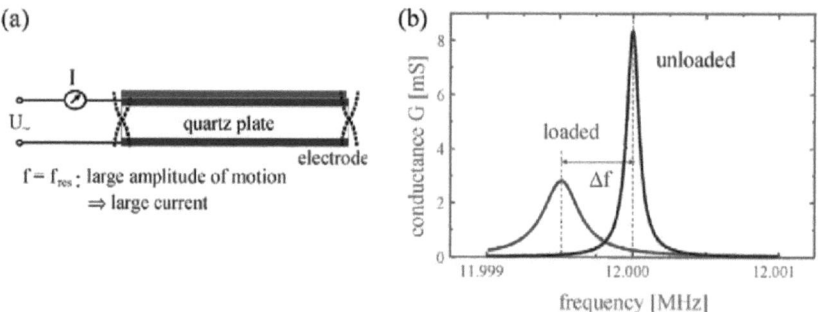

Figure 1: (a) Schematic diagram of the QCM and (b) the electrical characteristic with (loaded) and without (unloaded) adsorbed mass. Adopted from [27].

Surface Acoustic Wave (SAW) devices

A SAW biosensor, schematically shown in Figure 2, consists of one or more interdigital transducers (ITD) built on a piezoelectric substrate, such as quartz (α-SiO_2), lithium niobate ($LiNbO_3$), or lithium tantalite ($LiTaO_3$) [28]. The IDTs are interleaved electrodes that work as a transmitter to induce acoustic waves using an oscillating electrical potential and a receiver to convert acoustic waves back into an electrical signal. Between transmitter and receiver, the acoustic waves travel along the substrate, where the amplitude and velocity of the wave is influenced by adsorbed mass, viscoelastic changes and – to a far greater extent than in BAW devices - the conductivity of the surrounding liquid.

In addition to this rather simple set-up, the IDTs might be covered with a protective layer to avoid corrosion of the metal electrodes in buffer solution or the sensitive area can be covered with a layer with low acoustic velocity (e.g. a polymer [29] or SiO_2 [30]) in order to trap the wave close to the surface and minimise the energy dispersed into the substrate or the liquid. The effect of trapping the energy in a layer with an acoustic velocity lower than the surrounding is called the Love wave effect. Another way to confine the acoustic energy near the surface is to use a half-wavelength mass grating; the resulting waves are called shear horizontal waves (SHW). While there were many different device types tested for usage in biosensors, so called surface transverse waves (STW) or Love waves, or a combination of both seem to be most promising for high performance sensors. An overview over the recent developments towards SAW biosensors can be found in [3].

The SAW devices can be structured using photolithography which allows the integration of a large number of sensors on a small area. Devices with fluidic volumes well below 1 µl have been developed [31]. SAW sensors have the highest theoretical mass sensitivity in terms of frequency shift per attached mass among the acoustic resonators [32] and with a shown limit of detection of lower than 0.08 ng/cm^2 [33] a robust sensor system based on SAW would be extremely competitive with existing commercially available technology. The drawback of the SAW sensors is that it is difficult to build robust devices because the frequency change is influenced by many factors such as the conductance of the liquid and the conductance, dielectric and elastic constants of the adsorbate [34]. These perturbations make quantitative measurement challenging.

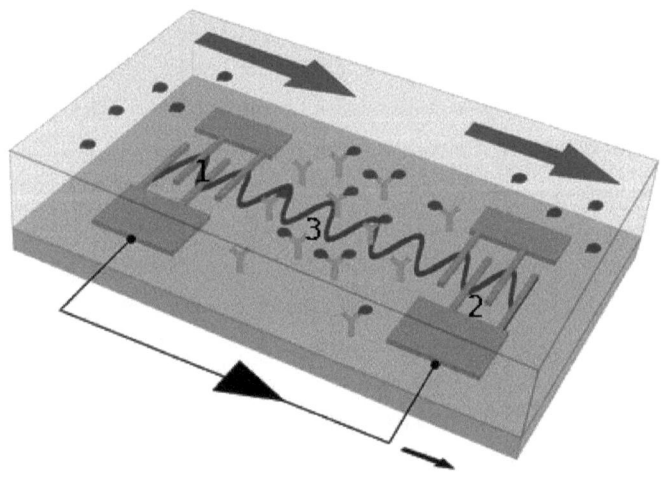

Figure 2: Typical set-up of a SAW biosensor: An acoustic wave propagates from a sender (1) to a receiver (2) passing the active sensor area (3) where its amplitude and velocity is influenced by the sensor surrounding (i.e. liquid or adsorbed mass) Adopted from [3].

Film Bulk Acoustic Resonator (FBAR)

FBARs are bulk acoustic wave (BAW) devices and, like the QCM, operate in the thickness shear mode (TSM). However, while the QCM has been used for decades for the analysis of intermolecular interactions, FBARs have only recently been produced for use in liquids [35-38], although thin film bulk acoustic resonators vibrating in longitudinal mode were previously fabricated for use in filters[39]. For applications in liquid, however, acoustic resonators operating in shear mode were developed, as the acoustic losses caused by longitudinal waves propagating into the liquid are too high to achieve sufficient Q-factors. Piezoelectric thin-films with the c-axis inclined from the film normal were developed to achieve sufficiently high piezoelectric shear coupling coefficients [40-47]. While the working principle of FBAR and QCM is similar, the QCM is produced in a top-down and FBAR in a bottom-up process using thin-film technology. As a result FBARs can be made thinner, which results in a higher resonant frequency. FBARs operating from some hundreds of MHz to several GHz have been presented [39]. However, determining the resonant frequency becomes more difficult for smaller devices, increasing the frequency noise. It was shown that the small size makes it possible to integrate many resonators on a small area [48]. This makes the FBAR especially promising for biomolecular interaction analysis with high throughput. The exact set-up of the FBARs used in this thesis is described in detail in Chapter 3.2.

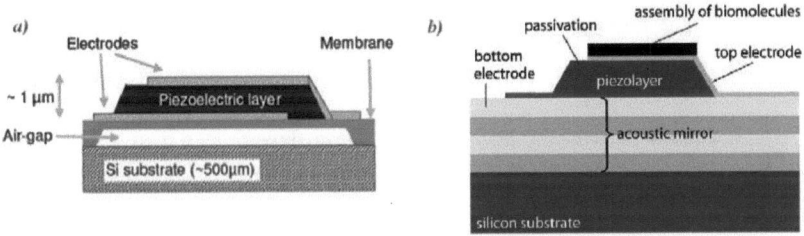

Figure 3: FBARs consist of a piezoelectric layer sandwiched between two electrodes over which the piezo layer is acoustically excited. The FBAR is isolated from the substrate by a) an air gap or b) an acoustic mirror. From [49] and [50].

1.1.2 Optical Sensors

The label-free optical biosensors introduced in this section are based on the interaction of light with the adsorbed biomolecules. Light is reflected at the active sensor's surface where it is affected by the amount of adsorbate present at the surface. The adsorption of biomolecules can be quantified in real-time by monitoring changes (e.g. intensity, wavelength, polarisation, and phase) in the light reflected at the active sensor surface.

Unlike acoustic sensors, most optical sensors are vicinity sensitive that means that substances do not need to be bound to the surface to be detected. Both substances bound to the sensor surface and substances that are close to the surface as well as changes in the optical properties (e.g. solvent concentration) of the liquid cause a signal. Bound and unbound substances can be distinguished using a reference channel with a passivated surface.

Surface Plasmon Resonance (SPR)

The SPR is the transducer with clearly the highest market share in the BIA market. This can be accredited to the high sensitivity of the technique [51], but also to the successful marketing concept of the leading vendor Biacore (GE Healthcare, Uppsala, Sweden) [52] and their large investments in the development of the technology [53] and especially the use of dextran matrix surfaces, which increase the effective surface area and thus the sensitivity [54].

Surface plasmons are oscillations of the free electron density in e.g. a metal. These plasmons can be excited when polarised light is diffracted from an interface between a dielectric and some metals at the angle of total reflection, gold being the most commonly used metal for biochemical measurements. The angle of total reflection depends on the refractive index of the surrounding medium within the decay length of the electromagnetic wave (called evanescent wave).

Figure 4 shows the set-up of an SPR sensor: The light emitted by a monochromatic light source is reflected at the interface between gold and liquid surrounding. The reflected light is detected and analysed. One way to readout the sensor signal is to measure the intensity of the reflected light for different angles. At the angle where the plasmons are excited, energy is adsorbed and the intensity of the reflected light has a minimum. This angle depends on the amount of mass adsorbed at the surface.

Rather than using a prism, as shown in the figure, the light can also be coupled in using an optical grating, and the wavelength or the intensity at a particular angle can be measured as an alternative to measuring the angle of minimum intensity. However, the combination of a prism coupler and minimum intensity angle readout is the most widely used as it has the highest sensitivity [55]. A more detailed overview of SPR technology can be found in [6].

Even though SPR requires a metal surface, many other functional layers can be put on top, e.g. the sensitivity-increasing carboxymethylated dextran surface introduced in 1990 by Löfås *et al.* [56].

One limitation of the SPR technology might be the substantial cost especially for systems with higher number of sensors usable in parallel like the Biacore 4000 with 20 individually accessible sensors in 4 different flow cells [57].

Since the signal change of the SPR induced by mass adsorption is reasonably well known (10 RU cm²/ng [58]), it is used in this thesis to determine the mass sensitivity of the FBAR (Chapter 5.2 and Chapter 7.2).

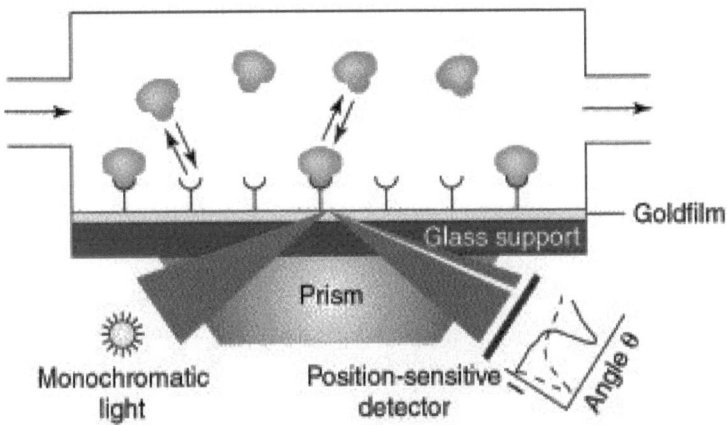

Figure 4: Schematic diagram of the SPR. Monochromatic light is reflected on a gold surface. At a certain angle, where the surface plasmons are excited, the reflected light has a minimum, which is continuously measured. This angle is directly connected with the analyte bound to the surface. From [1].

Surface Plasmon Resonance Imaging (SPRi)

The SPRi technology enables the building of microarrays based on SPR. In order to measure multiple sensitive spots using the same set-up, a CCD camera is used to record the intensity of the reflected light at a fixed incident angle and wavelength (Figure 5). Due to the higher complexity of this technique, the SPRi systems have a somewhat lower sensitivity than the SPR [59]. However, the published detection limit of 0.1 - 1 ng/cm² is sufficient for e.g. DNA [60] and protein [61] detection.

The number of parallel measurements in the literature is in the range of thousands but the possible number of sensors on an area of 1.4 cm² has been estimated to be more than

10'000 [62]. The number of sensitive spots is basically only limited by the available area and the number of individually accessible spots. While a high number of different substances can be easily immobilised by addressing single spots e.g. using a microspotter [63], it is difficult to access the functionalised spots with different ligand solutions.

This makes it easy to immobilise a high number of substances (e.g. proteins) and investigate their interaction with one or a small number of ligands (e.g. small molecules) but difficult the other way round. The fact that a wide range of measurements requires the immobilisation of few ligand targets and their testing against a high number of molecules as in drug discovery motivates developments towards the possibility of accessing a high number of pixels individually in flow [64, 65].

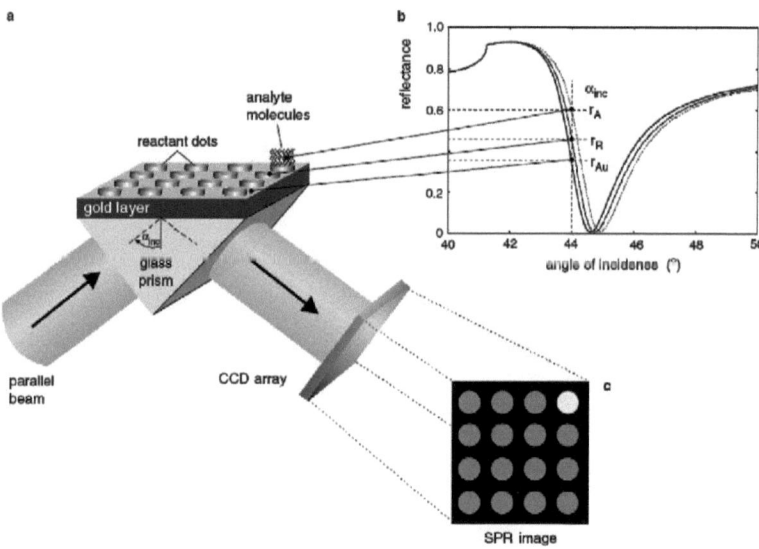

Figure 5: SPRi setup: Instead of measuring at one spot like with the SPR, the reflection of a number of spots is measured using a CCD camera. From [66]

Biolayer Interferometry (BLI)

BLI uses white light interferometry, a rather old technique commonly used to measure the thickness of transparent thin-films [67], to quantify the biomolecules adsorbed to the end of optical fibres. White light travelling through an optical fibre is reflected at the two surfaces: At the fibre-biomolecular layer interface and at the biomolecular layer – buffer interface. The reflected beams interfere generating a signal that depends directly on the number of adsorbed molecules [68].

The set-up using optical fibres makes an innovative sample delivery system possible: Instead of using a fluidic system to deliver the sample liquids to the sensor, the sensors (i.e. the optical fibres) are moved and dipped into well plates. A measurement sequence is performed by dipping the sensors into different reagent solutions. This makes the use of a fluidic system obsolete, which adds robustness to the systems and decreases maintenance and operating costs. Flow can be created, e.g. for diffusion limited reactions or to reduce rebinding when measuring off rates, by shaking the well in an orbital motion. Up to 16 sensors can be used in parallel by the Octet system (ForteBio, Menlo Park, CA). Because only substances bound to the sensor surface are detected, there is little influence from the media surrounding the sensor and thus no reference channel is needed. The downside of the BLI might be the low mass sensitivity of around 0.1 ng/cm^2, which makes it difficult to follow the adsorption of small molecules [69].

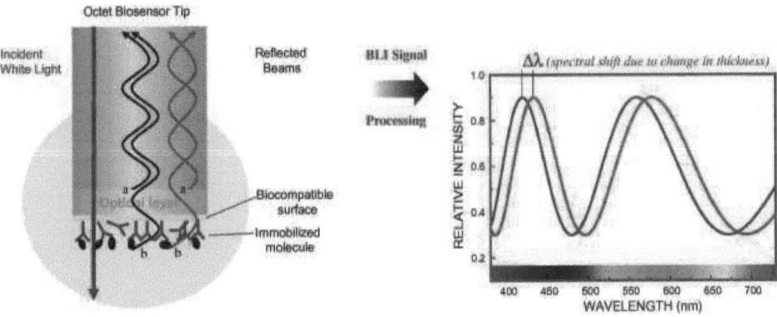

Figure 6: An optical fibre used for Bio-Layer Interferometry and a typical optical signal. From [68].

Diffraction grating based sensors

Diffraction grating based sensors measure the reflection of light from a photonic crystal. A photonic crystal is an optically regular structure made of a dielectric material, e.g. a grating comprising holes and spaces in the nanometre dimension. These gratings have been developed for use as biosensors [70]. When white light is radiated onto the grating, light of only a single wavelength is reflected. The wavelength of this light changes when biomolecules adsorb to the surface of the photonic crystal. For this type of photonic crystal a detection limit for protein of around 0.1 ng/cm² has been reported [11].

The advantage of this technology lies in the cheap manufacturing process and the resulting possibility of building a highly multiplexed sensor. SRU Biosystems, Inc. (http://www.srubiosystems.com) commercialised this technology under the name BIND™. They provide the sensors in microplates with 96-, 384- and 1536-well formats.

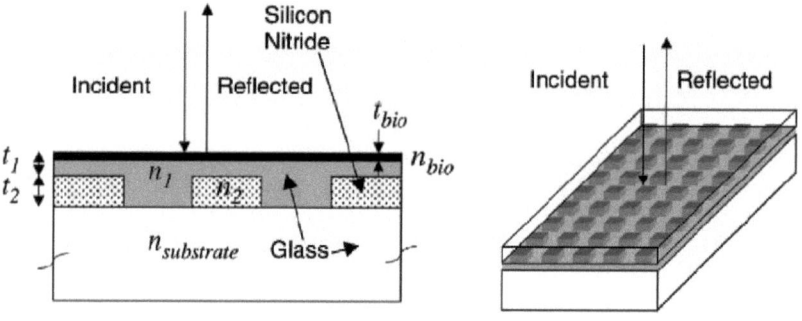

Figure 7: Schematic diagram of the photonic crystal used for colorimetric resonant reflection. From [70].

Optical-waveguide-based transducers

Optical-waveguide-based biosensors couple light into an optical waveguide. When the incident beam fulfils the condition of total reflection, the light forms a standing wave inside the waveguide, whose eigenvalues depend on the boundary conditions of the waveguide.

The intensity of the coupled light depends on the refractive index and the thickness of the layer of biomolecules adsorbed on the surface of the waveguide [71, 72]. This waveguide has to be transparent with a refractive index higher than the surrounding media and the thickness has to be of the order of the wavelength of the incident light. Dielectric metal oxides (TiO_2, Ta_2O_5, SiO_2, ZrO_2, Nb_2O_5) have been used as coatings because of their high refractive index and because they are corrosion resistant in buffer solutions. With the use of a conductive coating such as indium doped tin oxide (ITO) optical-waveguide-based biosensors can be combined with an electrochemical sensor which increases the spectrum of possible applications of this technology [73].

There are a range of different optical-waveguide-based biosensors that differ in the way the light is coupled into the waveguide and the way the coupled light is detected: The light can be coupled into the waveguide using an optical grating, or by putting the light source directly in line with the wave guide. Similarly, coupled light can be guided from the waveguide to the detector using a grating or directly. As an example, with Optical Waveguide Lightmode Spectroscopy (OWLS) the light is coupled into the waveguide using a grating and is detected directly (Figure 8). A comprehensive review about theory, methods and applications can be found in [5].

If the refractive index of the adsorbed material is known the thickness of the adsorbed material can be calculated. Otherwise, measuring both the transverse electric (TE) and transverse magnetic (TM) modes is required in order to calculate the refractive index and the thickness. While the capability of measuring the refractive index and the film thickness at the same time is an advantage, the sensitivity might be the main disadvantage. The limit of detection of OWLS has been reported to be 0.5 ng/cm^2 [74].

Dual-Polarisation Interferometry (DPI) is a technique in which light is introduced into two waveguides, one of which is for reference without liquid contact, the other is in contact with the liquid surrounding. After exiting the waveguide, the light is from the two waveguides is allowed to interfere. As one of the light beams has undergone a phase shift due to the contact with the liquid surrounding, the number of adsorbed biomolecules can be determined from the interference pattern [75, 76].

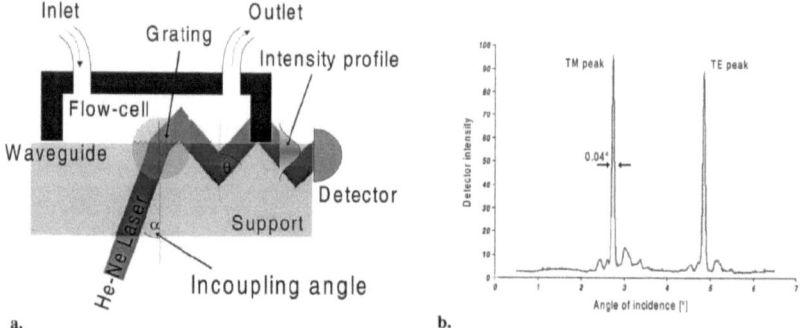

Figure 8: Working principle of the OWLS: (a) Light is coupled into an optical waveguide via an optical grating and the intensity is measured as a function of the incident angle. From the two peaks in the intensity spectrum (incoupling angles) (b), the thickness and the refractive index of the adsorbed layer can be calculated. (From [74])

Ellipsometry (ELM)

Ellipsometry (ELM) is a technique that measures the changes in the state of polarisation of elliptically polarised light, which is reflected at planar surfaces [77]. If the available measurement data is very accurate, both the refractive index and the thickness of the adsorbed layer can be obtained from the changes in the ellipsometric angles [78]. Assuming that the refractive index of protein films is around 1.5 the film thickness can be calculated more easily [74]. The complex theory behind the calculations, especially if systems with unknown optical properties are investigated, together with the requirement for reflecting surfaces are the main disadvantages of this technique. Imaging ellipsometry has been reported to allow measuring more than 10^5 pixels on an area of less than one cm^2 in one second. For this technique a CCD camera was used as a detector. ELM allows determination of the thickness of solid thin-films in air with accuracy well below 1 Angstrom, the detection limit for the adsorption of biomolecules is average: A detection limit of around 1 ng/cm^2 has been reported for surface plasmon enhanced ellipsometry [79].

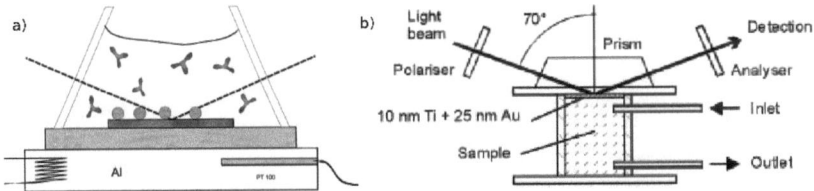

Figure 9: Set-up of the ELM (a) and the surface plasmon enhanced ELM (b) From[80] and[79].

1.1.3 Isothermal Titration Calorimetry (ITC)

In isothermal titration calorimetry (ITC) a solution of one type of biomolecule is titrated into the solution of a binding partner and the heat adsorbed or generated by the biochemical reaction is measured. From the heat of reaction for different concentrations the binding constant, K, the number of binding sites or the stoichiometry (n) and thermodynamic data, the enthalpy ($\Delta H°$) and entropy (ΔS) of the binding, can be determined in a single measurement.

Being able to measure heat effects as small as 0.4 µJ (0.1 µcal) allows determination of binding constants, K's, as large as 10^8 to 10^9 M^{-1}. The typical setup consists of a sample and a reference cell in a thermostatically controlled environment, a syringe to introduce the ligand solution into the sample cell, a means to keep the sample cell at the same temperature as the reference cell and to measure the heat changes.

The cell volume is typically in the ml range and the injected volume can vary from about 1 to 20 µL but usually at relatively high concentrations [81].

The large number of parameters that can be measured at the same time and the fact that the reaction can be performed in solution and neither a label nor the immobilisation on a surface is required are unique features of this technique. The great experimental effort in planning and performing the measurement and the high sample consumption (generally some grams of the protein samle are required) are the drawbacks of this technique.

Differential scanning calorimetry (DSC) is a related technique, in which the temperature of a biomolecular solution is changed and the resulting heat change is measured. This gives information about e.g. conformational changes of proteins.

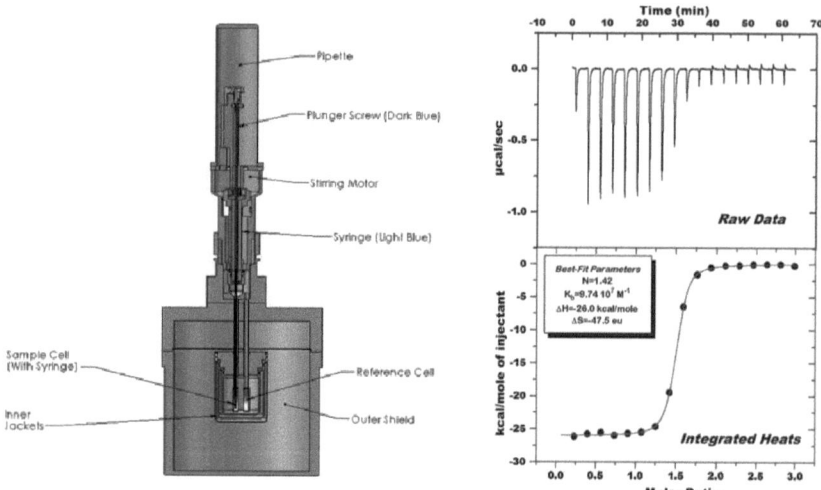

Figure 10: a) ITC setup and b) typical measurement curve: the raw data and the isothermal (From www.microcal.com)

1.1.4 Electrochemical Sensors

Label-free electrochemical sensors are based on measuring the change in charge, current, potential or conductivity that takes place when the target molecule binds to the functionalised sensor surface.

Another technique is measurement of the conductivity of the solution. Many reactions produce or consume electrons thereby altering the overall electrical conductivity of the solution, and this mechanism can be exploited. However, because the conductivity of a solution depends on all present ions, this sensing principle is considered to be rather non-specific. Amperometric biosensors measure a change in current. However, many biomolecules like proteins are not electroactive, and therefore require labelling [82].

Impedance sensors measure the electrical impedance between an electrode and the solution at a fixed or variable frequency. The latter approach is called electrochemical impedance spectroscopy (EIS). Upon adsorption of the target molecule to the electrode, the impedance undergoes a detectable change, which has been shown for a variety of chemical systems [83]. In cyclic voltammetry (CV), the applied voltage is slowly changed and the resulting current is measured. A change in current represents a change of electron transfer resistance using a redox couple such as ferri/ferrocyanide. Molecules adsorbed to the surface act as insulator and increase the resistance Figure 11 a) shows CV curves for a bare gold electrode (a), the adsorption of protein A (b) and IgG (c) at a scan rate of 50 mV/s. Figure 11 b) shoes the corresponding EIS measurement. Plotted is the real part versus the imaginary part of the electrical impedance from a frequency range from 100 kHz to 0.1 Hz.

Electrochemical sensors based on field effect transistors (FET) consist of a transistor where the metal gate is replaced with an appropriate functionalisation. Upon adsorption of the target molecule, the potential at the gate oxide changes resulting in a measurable signal between source and drain[84]. One hindrance to commercial success of FET based biosensors apart from the high cross-sensitivity e.g. to changes in pH might be the unsolved challenge of the incorporation of a high quality but economic reference electrode [82].

Because the sensitivity depends strongly on the surface chemistry the limit of detection cannot be expressed in terms of ng/cm^2.

Electrochemical sensors can be combined with other label-free transducers by integrating a conductive electrode to the setup. This has been shown e.g. for OWLS [85, 86], SPR [87, 88], ELM [89-91], QCM-D [92, 93]. These combined set-ups enable measurement of the adsorption under applied electric field or simultaneous measurement of the adsorption and electrochemically analysis of the adsorbate that provides additional information about the character of biomolecular interactions.

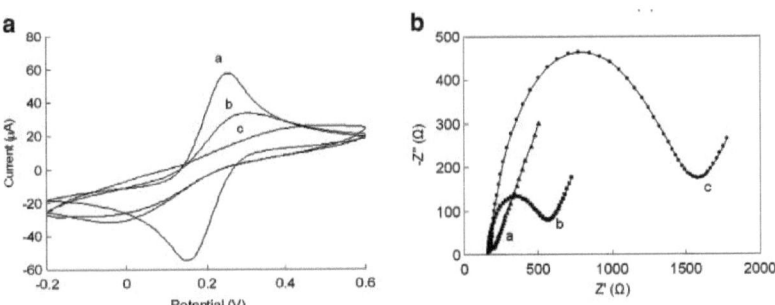

Figure 11: Examples for a measurement curve of CV (a) and EIS (b). The adsorption of molecules to the surface can be seen from an decrease in current (CV) and an increase in impedance (EIS) From [94].

1.1.5 Nanostructure Biosensors

Feynman's "Plenty of Room at the Bottom" [95] might prove equally valid for label-free biosensors. There are nanostructured biosensors emerging that are based on transducers with at least one dimension of the biosensor at the nanoscale. Along with lower detection limits come lower liquid and sample consumption and the potential to integrate a large number of sensors within a small space or area.

Nanoplasmonics

While conventional SPR uses surface plasmons excited at the interface between a dielectric and a macroscopic gold film, localized surface plasmons resonance (LSPR) can be excited in noble metal nanostructures. These nanostructures can be particles like disks [96], rings [97] or nanoholes in thin-films [98]. They can also be arranged in lines [99] or arrays [100]. A peak in the adsorption spectrum appears if the wavelength of the incident photon is resonant with the localized surface plasmons of conduction electrons of the nanostructure [101]. The position and height of this peak depends on the size, shape and composition of the nanostructure and the local dielectric environment [102, 103]. The latter enables the measurement of the adsorption of molecules on or inbetween the nanostructures.

The small size of the particles allows detection of very low quantities of adsorbate. The combination of detection limits that are comparable to current commercial instruments together with the small size of the particles makes single molecule detection probable [96]. It is important to note that while single molecule sensitivity sounds like a powerful property, commercial application might be difficult because of the time required to scan a sample liquid for single molecules.

As well as the sensitivity in terms of signal-to-noise ratio, which is similar to commercially available SPR systems [104], nanoplasmonic sensors offer other advantages: the sensitivity to bulk refractive index changes is more than one order of magnitude lower, which might make temperature stabilisation obsolete and increase the resiliance to small changes in organic solvent (e.g. DMSO in the drug discovery process [105]). Additionally, the required set-up is simpler than for the SPR as the light can be irradiated by a white light source at any angle and therefore does not need any prism for coupling [9]. For a successful commercialisation the challenge of producing the required nanostructures in a cheap, robust and reproducible way has to be overcome. Furthermore, the sensing length of nanopasmonics is only in the range of few nanometres [98], which might too short for many applications.

Nanowire Biosensors

Nanowire biosensors are mainly employed as miniaturised electrochemical sensors. Biochemically functionalised they can be used in AC voltammetry [106] or function as gate in FETs [107] amongst others by connecting them between source and drain. As with electrochemical sensors using thin-films, adsorbed biomolecules change the dielectric environment around the nanowire. Due to the small size of the nanowires and the resulting small surface-to-volume ratio, biomolecules binding to the nanowire result in a significant change in the electrical properties of the nanowire [12]. Increasing sensitivity with decreasing nanowire diameter has been shown both in theory and experimentally [108]. With the diameters being comparable to the size of the biochemical analytes under analysis [8], extremely high sensitivities down to the detection of a single virus have been shown [109]. Multiplexed detection of proteins was also demonstrated on a multiplexed nanowire sensor [110]. However, the sensitivity is significantly reduced in solutions with high ion concentrations when the distance between the analyte to the nanowire after adsorbtion is greater than the Debye length, due to the charge of the analyte being shielded by the ions in solution [111]. Three possible materials investigated for use as nanowires are carbon nanotubes (CNTs), silicon nanowires (SiNWs) and conducting polymer nanowires (CP NWs). CNTs are interesting because they are mechanically stable and exist, dependent on their structure, both as semiconductor and conductor, so that they could be used for several parts of the FET and the connectors. SiNWs and CNTs have a high tensile strength and Young's Modulus, however, they are always semiconducting. Both the CNTs and the SiNWs are produced under harsh conditions, so the biochemical functionalisation has to be done after production. This is different for CP NWs, which can be synthesized under ambient conditions using well-known chemical processes and therefore can be functionalised before or during synthesis. A variety of techniques have been employed to assemble the nanowires into functional devices including alignment in electric and magnetic field, lithography, Langmuir – Blodgett techniques and biomolecule mediated self-assembly [112].

However, the assembly and the precise and reproducible alignment of the nanowires, the interconnections with the read-out electronics together with the production with controlled dimensions (e.g. length and diameter), physical properties, and high purity in large-scale manufacturing processes at reasonable costs are all challenges waiting to be overcome to achieve commercial success [8, 12, 112].

	Company Name	Product Name	Technology	Limit of Detection [ng/cm^2]	Number of Parallel Sensors	Overall Sample Volume[5]	Sample Volume per Sensor/Pixel[6]	Web Address	Comments
SPR[1]	GE Healthcare	Biacore 4000	optical	0.01	16	60 µl (For 4 flow cells)	4 µl	www.biacore.com	
		Biacore T100		0.01	4	20 to 50 µl	21 to 50 µl		
SPRi[2]	Horiba	SPRi-Plex™	optical	0.5	up to 1000	1.6 ml	2 µl (target) / 1.6 ml (ligand)	www.horiba.com	Up to 1000 substances can be spotted, only one substance can be measured in flow
BLI	ForteBio	Octet RED384	optical	0.1	16	n/a	200 µl	www.fortebio.com	
Diffraction Grating Based[3]	SRU Biosystems	BIND	optical	0.01	96-, 384- and 1536-well microplate	n/a	down to 5 µl	www.srubiosystems.com	
	Corning	Epic		0.5	384-well microplate	n/a	15 - 30 µl typical	www.corning.com	
Optical Waveguide Based	MicroVacuum Ltd.	OWLS 210	optical	0.5	1	n/a	20 to 250 µl	www.owls-sensors.com	
	Farfield	AnaLight 4D		0.01	1	n/a	50 µl	www.farfield-group.com	
ELM	Maven Biotechnologies	LFIRE	optical	0.1	1	n/a	n/a	www.mavenbiotech.com	
QCM[4]	Q-Sense	E4 Auto	acoustic	0.5	4	n/a	400 µl	www.q-sense.com	
SAW	SAW instruments GmbH	sam5	acoustic	0.05	5	40 to 80 µl	8 to 16 µl	www.saw-instruments.de	
FBAR	n/a	n/a	acoustic	0.5	up to 64	100 µl	2 µl	n/a	
Electrochemical	Eco Chemie	n/a		n/a	n/a	n/a	n/a	www.ecochemie.nl	
ITC	MicroCal	iTC$_{200}$	calorimetric	n/a	1	n/a	n/a (at least 10 µg protein)	www.microcal.com	in-solution, no immobilisation needed

Table 1: Overview of the commercially transducer systems for BIA together with the FBAR.

1 Other SPR systems: Bio-Rad ProteOn XPR36 (www.bio-rad.com), Eco Chemie Autolab TWINGLE (www.ecochemie.nl), Reichert Inc. SR7000DC (www.reichertspr.com)
2 Other diffraction grating based systems: Axela dotLab (www.axelabiosensors.com),
3 Other SPRi systems: Biacore Flexchip (discontinued), Plexera Bioscience PlexArray™ (www.plexera.com), GWC Technologies SPRimager®II (www.gwctechnologies.com), IBIS Technologies IBIS-ISPR (www.ibis-spr.nl)
4 Other QCM systems: Sierra Sensors QCMA-1 (www.sierrasensors.com), TTP LabTech RAP (www.ttplabtech.com), Attana A200(www.attana.com)
5 Sample volume means the minimum of volume required to follow one binding interaction.
6 The sample volume per pixel can vary from the overall sample volume if more than one pixel is in one flow cell.

1.2 Label-free Transducer Principles to Investigate Conformational Changes

While this thesis deals mostly with FBAR measurements where the mass of the adsorbate is measured in order to obtain quantitative information, measurements like the ones with calmodulin in Chapter 8 show the potential of monitoring other measures in addition to the adsorbed mass in order to detect conformational changes e.g. in proteins.

Measuring label-free conformational changes is less common than measuring the adsorption of proteins. However, because it gives additional information to compare to the asorption this technique might be of interest for use in drug discovery.

There are many technologies available that use labelling (e.g. NMR or X-RAY crystallography) to show conformational changes, however, this overview concentrates on label-free technologies. Label-free technologies generally give a lower information quality when compared to techniques such as NMR, which can measure the positions of atoms in a molecule with high accuracy. However, their sample consumption and the preparatory effort required is much lower, and the throughput potentially much higher.

In this overview each technology is briefly described and sample measurements for calcium binding to calmodulin are given because the calcium/calmodulin is the model used on the FBAR in Chapter 8.

1.2.1 QCM/QCM-D

The QCM for application in BIA as a mass sensitive device is described in Chapter 1.1.1. If in addition to the resonant frequency, as a measure for the dissipated acoustic energy, the motional resistance [21], the conductance [22], the Q-factor or the energy dissipation [23] are measured, information about the conformation of proteins can be obtained. Measurements yielding observations of cross-linking between proteins [26], or the folding of Calmodulin upon calcium binding have already been published [21, 22]. An example measurement for the Ca/Calmodulin system is shown in Figure 12.

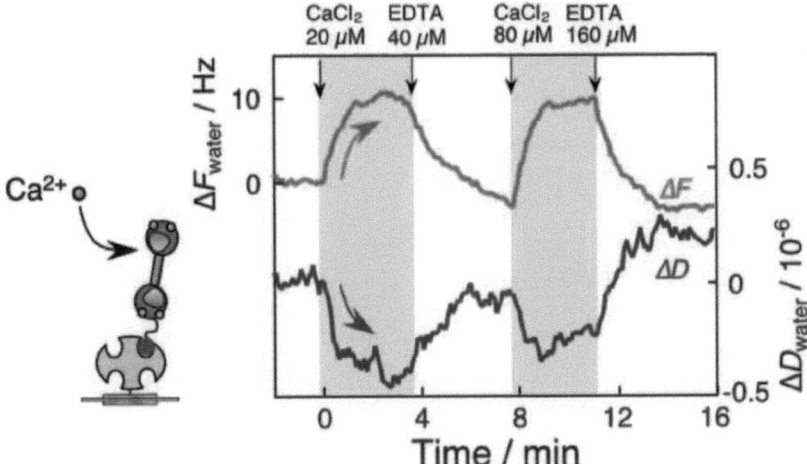

Figure 12: Frequency shift and dissipated energy on QCM. The folding of the Calmodulin upon addition of calcium can be seen from an increase in the resonant frequency and a decrease in dissipation. Addition of EDTA removes the calcium form the system and the process is reversed. From [22].

As an acoustic technology measuring acoustic properties of the adsorbate, information about coupled water, stiffness and viscosity can be obtained. The sample consumption is in the 100 µl range per measured protein.

1.2.2 Dual Polarisation Interferometry (DPI)

Dual Polarisation Interferometry (DPI) is a waveguide based biosensor and is described previously in the section about optical sensors (see page 7). The high sensitivity of the technique allows simultaneous measurment of both thickness and refractive index of the adsorbate and thus can detect conformational changes if the conformational change involves a change in refractive index. Figure 13 shows the folding of calmodulin caused by calcium binding on the DPI *Ana*Light® instrument (Farfield sensors Ltd, Manchester, UK). The Farfield group itself provided the measurement data.

(a)

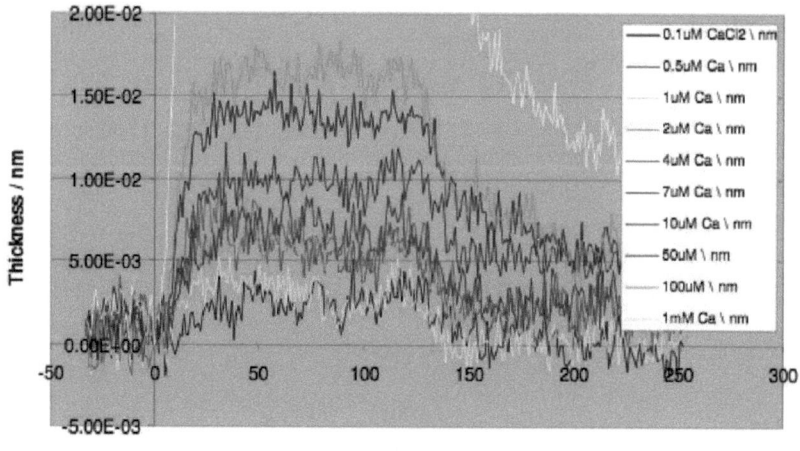

(b)

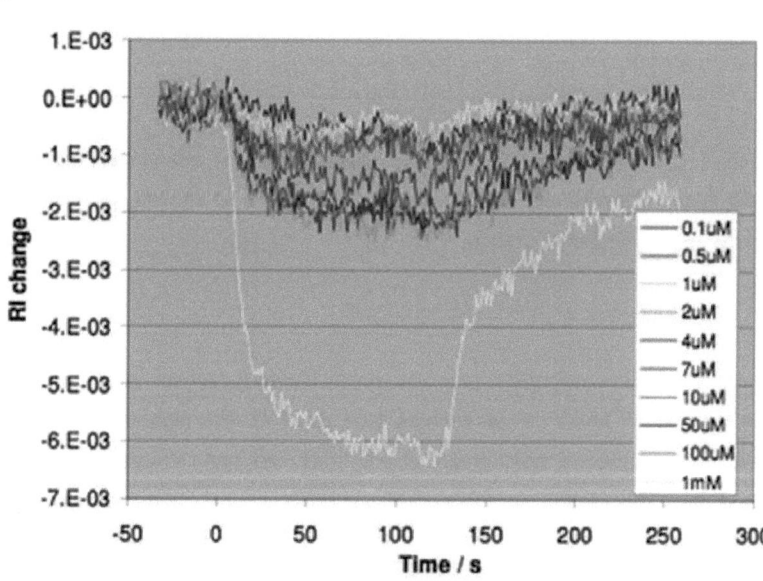

Figure 13: Calmodulin folding monitored in real-time on DPI. Diagram (a) shows the changes in thickness for different calcium concentrations. Diagram (b) shows the change in refractive index for different calcium concentrations. Measurement data is courtesy to the Farfield group.

1.2.3 Backscattering Interferometry (BSI)

The use of backscattering interferometry (BSI) for the detection of conformational changes was presented recently and is not yet commercially available [113]. BSI is an optical method that measures the fringe pattern of LASER light scattered on the sensor cell (Figure 14a). Refractive index changes can be seen in the fringe pattern, which then allows conclusions to be drawn about the conformational state of molecules.

The unique feature of BSI is the fact that this technique is solution based, meaning that the target does not have to be immobilised on a surface of the sensor but the binding can take place in solution. It is also compatible with microfluidics, which results in a very low sample volume. Sample consumption as low as 4 µl has been shown in literature [113].

Figure 14b shows data of a measurement where calcium in various concentrations in the micromolar range binds to calmodulin. The diagram shows that the advantage of this technique is not only the absence of the requirement to immobilise samples on a surface and the low sample volume needed but also a very good signal-to-noise ratio.

(a)

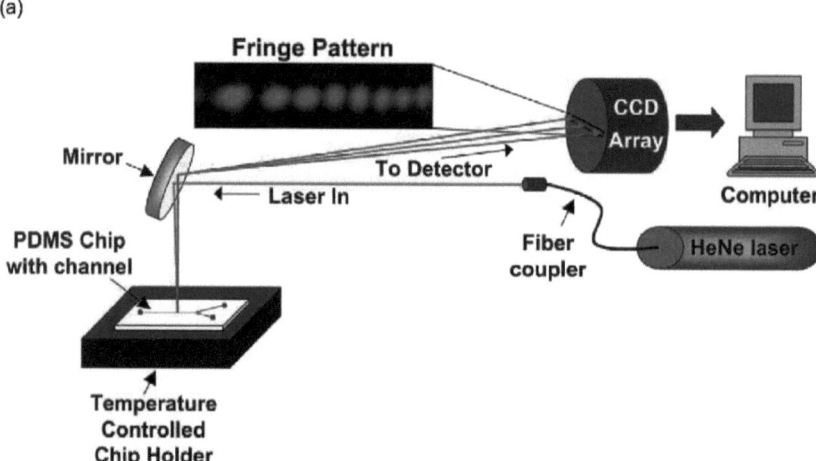

(b)

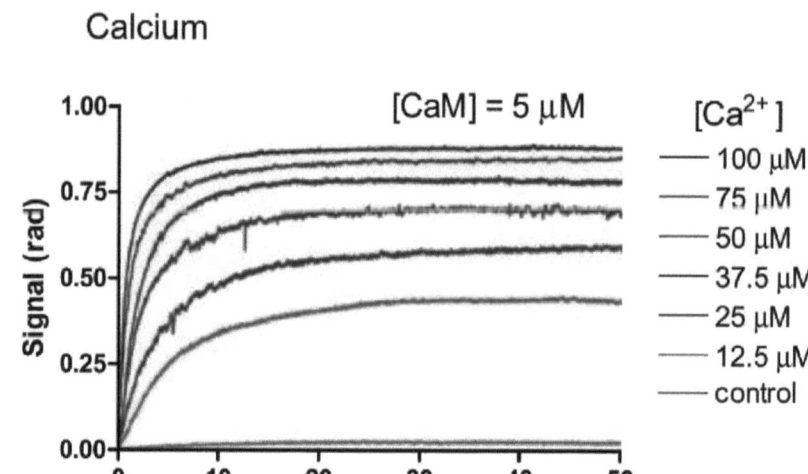

Figure 14: backscattering Interferometry: (a) shows the set-up. A LASER beam irradiates the sample volume. The pattern of the scattered light allows drawing conclusions about the refractive index of the solution. (b) shows an example measurement of different concentrations of calcium binding to calmodulin.

1.2.4 Isothermal Calorimetry (ITC) and Differential Scanning Calorimetry (DSC)

Isothermal calorimetry (ITC) and differential scanning calorimetry (DSC) are well established methods [114, 115]. ITC was described previously in Chapter 1.1.3. DSC is a related technique with the difference that the heat required to increase the temperature of the sample and a reference with known heat capacity is measured [116].
Both techniques are solution-based like the BSI, however, the sample consumption and the experimental effort is much greater.

Figure 15 shows DSC and ITC measurements investigating calcium binding to calmodulin (From [117]).

(a)
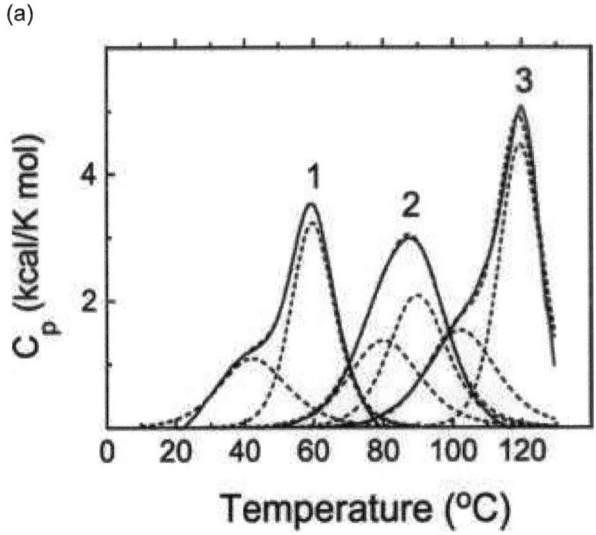

(b)

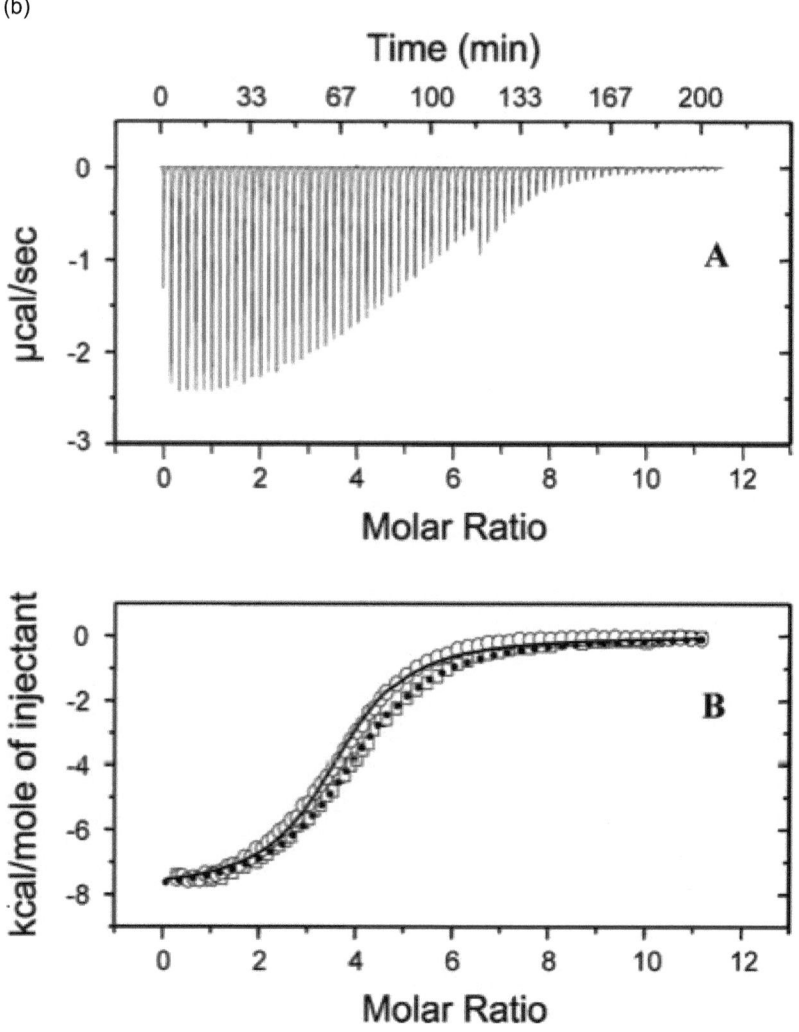

Figure 15: ITC and DSC measurements investigating calmodulin: (a) shows the heat capacity depending on the temperature for (1) apocalmodulin, (2) oxidised Ca^{2+} calmodulin and Ca^{2+}-calmodulin. (b) shows the ITC measurements of calcium binding to calmodulin.

	QCM/QCM-D	FBAR	DPI	BSI	Calorimetry
Vendor name	Q-Sense	n/a	Farfield	n/a	GE Healthcare
Product name	E4	n/a	Analight	n/a	Microcal
Technology	acoustic	acoustic	optical	optical	calorimetric
Measured properties	adsorbed mass and dissipation	adsorbed mass and dissipation	thickness and refractive index	refractive index	adsorbed heat, enthalpy
Number of parallel sensors	4	64	1	1	1
Sample volume	100 µl	3 µl per pixel	< 100 µl	4 µl	200 µl
Amount of protein needed	100 pmole	50 fmole per pixel	n/a	pmole range	nmole range
Required protein concentration	µM range	µM range	n/a	µM range	mM range
Volume of analyte needed	100 µl	100 µl	< 100 µl	4 µl	µl range
Limit of detection for Calcium binding to calmodulin	µM	µM	µM	µM	< µM
Web address / source	www.q-sense.com / [21, 22, 118]	Chapter 8 of this thesis	www.farfield-group.com / Chapter 1.2.2 of this thesis	[113]	www.microcal.com / [117]
Comments				In-Solution	In-Solution

Table 2: Overview of label-free technologies for investigation of conformational changes of biomolecules.

1.3 Conclusion and Outlook

A broad variety of transducer principles for BIA have been introduced. These included optical (SPR, SPRi, BLI, diffraction grating based sensors, waveguide-based sensors, ELM), acoustic (QCM, SAW and FBAR), electrochemical and calorimetric sensors. Their different working principles result in different properties such as sensitivity, sample consumption or the potential for multiplexed sensing. While the sensitivities from literature were stated for all transducers, it should be kept in mind that the perceived sensitivity of the operator in daily routine use is critical for successful commercialisation. Factors other than the sensitivity play an important role in the formation of this perception, including usability, easy handling, reproducibility and robustness.

It is also important to take the state of development of the technology into account. The sensitivity of a sensor technology can change greatly over course of time and with more resourses for research and development. Biacore for example improved the surface mass sensitivity by a factor of 20 to 30 and the association constants by nearly 3 orders of magnitude in only one decade [119]. It can be therfore expected that sensors in early development stages might undergo a similar improvement.

In addition also other factors can play an important role. For example, while the FBAR might not be among the most sensitive devices, it shows interesting properties such as the sensitivity to viscoelastic properties that are investigated in Chapter 5.5, Chapter 6 and Chapter 8 of this thesis.

While the nanostructure-based sensors may have the most promising properties among the presented transducers in terms of sensitivity, sample consumption and number of parallel measurements, they also look to have the longest road to commercialisation in front of them. Apart from the potential toxicity of the nanoparticles [120] which have to be taken into account, the problem of producing them in an economic way in large quantities and high quality offers lots of interesting working tasks for research and development.

2 Thesis Scope

The aim of this thesis is to advance knowledge in the usage of FBARs in label-free monitoring of biomolecular interactions. This will consist of two main objectives:

1. Optimization of the FBAR design to improve performance for the application as a biosensor
2. Evaluatation of the FBAR behaviour in a variety of application-near measurements and compare them to other available transducers.

In the introduction (**Chapter 1**) a wide range of label-free transducers together with their most important properties like sensitivity were summarized. A small part of this section is dedicated to technologies that can be used to investigate conformational changes of proteins.

Chapter 3 describes the materials and methods used in this thesis.

Chapter 4 shows the results of measurements where materials with different acoustic properties are deposited on the FBAR and the response of the FBAR is monitored. It is well known that the FBAR can be used as a mass sensor when the frequency shift caused by the adsorption is linear to the absorbed mass. This is, however, only the case if the adsorbed layer is thin compared to the thickness of the resonator. In the measurements in this chapter, the film thicknesses were varied from a few nanometres up to few hundred nanometres.
With this range, films thicknesses small compared to, and of the order of the resonator thickness are covered.
The adsorption of materials was simulated for materials with different mass densities and acoustic velocities. Thin-films of platinum, aluminium oxide, tungsten, and carbon nanotubes were deposited on the FBAR and the results were fitted to the model used in the simulations.
The acoustic velocity of the carbon nanotube films was much lower than the other materials (Pt, W and Al_2O_3). Due to this interesting property, carbon nanotube thin-films are a promising material for acoustic devices where materials with particularly low acoustic impedance are desired.

In **Chapter 5** the FBAR is used as a biosensor. First the mass sensitivity of the FBAR was determined by performing a reference measurement parallel on FBAR, SPR and QCM. Having shown that the mass sensitivity is sufficient for detecting biomolecules, a series of measurements with the FBAR as a biosensor were performed. As a first example system the application of DNA detection is presented. For this application, the FBAR set-up comprising the FBAR, the read-out electronic, read-out software, the surface chemistry and the fluidic system had to be developed or improved. Especially important were the functionalisation of the individual FBAR pixels and the robustness of the overall system. The results show that the FBAR is capable of detecting DNA strands in buffer and diluted serum. Finally, measurements adsorbing S layer proteins to the sensor surface were performed. For the S-layer, a concentration dependent adsorption behaviour was found, and a crystallisation process could be distinguished from the adsorption process.

In **Chapter 6** two measurements were performed on the FBAR and compared with the results of the same measurements performed on the QCM: The adsorption of lipid vesicles followed by the formation of a lipid bilayer and the formation of a polyelectrolyte multilayer. As the basic physical principle of the two devices is similar, the main differences are the smaller size and the higher operating frequency of the FBAR. By performing the measurements on both systems, the influence of the differences on the results can be investigated. With both the vesicles and the polyelectrolyte multilayer, significant qualitative differences were found between the two technologies. Using these results, in combination with simulations, it was found that at the higher resonant frequency of the FBAR, viscoelastic properties and penetration depth significantly influence the FBAR's performance.

One of the interesting advantages the FBAR presents is the possibility of integrating a large number of sensors on a small area. Together with a high number of different functionalisations this makes this device interesting e.g. for point-of-care devices where a high number of substances should be detected in a body liquid using a small handheld device. **Chapter 7** presents the use of a multiplexed version of the FBAR for DNA detection. Using a CMOS-integrated FBAR array, 64 pixels were available for simultaneous measurement. This array was used to detect two different DNA sequences on one sensor. Measurements were performed in buffer and in diluted serum.

While the FBAR is used predominantly as a mass sensitive device in this thesis, in **Chapter 8** not only the mass of the adsorbate is detected but also the conformational state during the binding of calcium and the peptide CaMKII to calmodulin. Because the mass of the calcium is too small to be detected, the conformational change caused by the binding process is measured by monitoring the resonant frequency and the motional resistance of the FBAR. The resonant frequency is a measure of the amount of mass coupled to the sensor while the motional resistance is influenced by the viscoelastic properties of the adsorbate.

The measured frequency shift during the calcium adsorptions was found to be highly dependent on the surface concentration of the immobilized calmodulin, which indicates that the measured signal is highly influenced by the amount of water inside the calmodulin layer.

By plotting the measured motional resistance against frequency shift, a mass adsorption can be distinguished from processes involving measurable conformational changes. Using this method, three serial processes could be identified during the peptide binding. The conclusion of this chapter is that the FBAR is a promising technology for the label-free measurement of conformational changes.

Chapter 9 contains the conclusion of the thesis. The results of the measurement are summarised and future development possibilities for the FBAR technology are presented.

3 Materials and Methods

3.1 Chemicals, Proteins, Polymers & DNA

Ultrapure water (Milli-Q gradient A 10 system, with a resistance of 18.2 MΩ·cm, < 4 ppb, Millipore Corporation, USA) was used for the preparation of all solutions. All the chemicals used were of analytical grade.

Polydimethylsiloxane (PDMS) was purchased from Dow Corning (Sylgard 184, USA). To prepare, it was mixed at a 10:1 ratio with the curing agent and degassed in a vacuum desiccator. After casting into a mould, the PDMS was cured at 80°C for a minimum of four hours.

Tris(hydroxymethyl)aminomethane, 6-mercapto-1-hexanol (MCH), bovine serum albumin (BSA), minimum 98% purity, were purchased from Sigma Aldrich Finland Oy (Helsinki, Finland). N,N-bis(2-hydroxyethyl)-a-lipoamide (Lipa-DEA) was prepared as previously described [121]. Hydrogen peroxide (30 %) and the solvents EDTA and Na_2HPO_4 were purchased from Merck KGaA, sodium chloride and NaH_2PO_4 from J. T. Baker. Sodium hydroxide (NaOH) was purchased from AkzoNobel (Sweden) and EDTA (Ethylenediaminetetraacetic acid) from Fluka AG (Buchs, Germany). Ammonium hydroxide (28-30 % NH_3), bovine serum albumin (BSA, minimum 98 % purity), sodium dodecyl sulphate (SDS) and human serum from male AB plasma (H-4522) were purchased from SIGMA (Mannheim, Germany). Human serum was diluted at a ratio of 1:100 in PBS running buffer (20 mM Na_2HPO_4, 0.3M NaCl, 1 mM EDTA pH 7.4) and then filtrated through a 0.45 µm syringe filter (PALL Corporation, Cornwall, UK) prior to use.

The DNA probes were selected due to their relevance in breast cancer diagnosis [122] and custom synthesized and HPLC purified by Metabion (Martinsried, Germany). A phosphate-buffered saline (PBS) buffer solution of 20mM Na_2HPO_4/NaH_2PO_4, 300mM NaCl, 1mM EDTA, pH 7.5 was used in all measurements.

The disulphide modified single-stranded (ss) DNA-probes, which were assembled on the gold surfaces were: DMT-S-S-PTGS2-27 with sequence of 5´-DMT-S-S-C6-CGA TTG TAT TCG GAT AGG ATT TTA TGG-3´ and DMT-S-S-CALCA-25: 5´-DMT-S-S-C6-GCT TCC GAT CAC ACT CAT TTA CAC A-3´. DMT-S-S- refers to a disulphide modification of the surface oligos. C6 refers to a spacer of 6 C base pairs (CCCCCC). The sequences for the complimentary ss-oligos were: PTGS2-27: 5´-CCA TAA AAT CCT ATC CGA ATA CAA TCG-3´ and CALCA-25: 5´-TGT GTA AAT GAG TGT GAT CGG AAG C-3´. Both the complimentary ss-oligos and the longer ss-strands (PTGS2-123 and CALCA-92) were used as analytes. The sequences for long synthetic ss-strands, along with the hybridization sites (underlined) for complimentary oligos attached on the surface are presented here: PTGS2-123: 5´-CTA TAT CCA ACC CCA CTC CTA ATA AAA CAA CCA AAA AAC AAA CTT ACG TAT TAA ACA ATT TTC TCC ATA AAA TCC TAT CCG AAT ACA ATC GCA CTT ATA CTA ATC AAA TCC CAC ACT CAT ACA-3´ and CALCA-92 respectively: 5´-TGG GTA TAT GTT GGG AGA TAG TAA TGG GTT TGG GTG TGT GTA AAT GAG TGT GAT CGG AAG CGA GTG TGA GTT TGA TTT AGG TAG GGA TTA TA-3´. All DNAs used in this thesis are summarised in Table 3.

The DNA products were solubilised in MilliQ water and diluted in phosphate running buffer at desired concentrations just prior to use. N,N-bis(2-hydroxyethyl)-α-lipoamide (Lipa-DEA) was prepared as previously described [23].

Name	Sequence (5' - 3')	5'	3'	Mer
DMT-S-S-PTGS2-27	CGA TTG TAT TCG GAT AGG ATT TTA TGG	DMT-S-S-C6	-	27
DMT-S-S-CALCA-25	GCT TCC GAT CAC ACT CAT TTA CAC A	DMT-S-S-C6	-	25
PTGS2-27	CCA TAA AAT CCT ATC CGA ATA CAA TCG	-	-	27
CALCA-25	TGT GTA AAT GAG TGT GAT CGG AAG C	-	-	25
PTGS2-123	CTA TAT CCA ACC CCA CTC CTA ATA AAA CAA CCA AAA AAC AAA CTT ACG TAT TAA ACA ATT TTC TCC ATA AAA TCC TAT CCG AAT ACA ATC GCA CTT ATA CTA ATC AAA TCC CAC ACT CAT ACA	-	-	123
CALCA-92	TGG GTA TAT GTT GGG AGA TAG TAA TGG GTT TGG GTG TGT GTA AAT GAG TGT GAT CGG AAG CGA GTG TGA GTT TGA TTT AGG TAG GGA TTA TA	-	-	92

Table 3: DNA sequences used in this thesis

The strand PTGS2-27 is complementary to DMT-S-S-PTGS2-27, while CALCA-25 is complementary to DMT-S-S-CALCA-25. The strand PTGS2-123 contains a region complementary to DMT-S-S-PTGS2-27. The strand CALCA-92 contains a region complementary to DMT-S-S-CALCA-25.

3.2 The FBAR and the Measurement Set-up

The FBARs used in this thesis consist of a 500 nm thick piezoelectric zinc oxide layer sandwiched between a 100 nm aluminium top electrode and an 890 nm tungsten bottom electrode. In Chapter 4 a simpler set-up is used which is described in the experimental section in Chapter 4.1.1. The ZnO was deposited by reactive sputtering using a blind system which is described in detail in [123]. This deposition process formed a ZnO layer with the c-axis inclined against the layer normal resulting in effective shear coupling coefficients of up to 19.2 %. The resonators were covered with a 300 nm thick SiO_2 layer to isolate the electrodes from the liquid and mounted on a quarter-wavelength acoustic mirror (made of SiO_2 and W) in order to avoid the propogation of acoustic waves into the substrate. This configuration using an acoustic mirror is called surface mounted resonator (SMR). The lateral size of one sensor surface was 200 µm x 200 µm. The gap between two adjacent resonators was 100 µm. The resonators could be either used with the SiO_2layer in contact with the liquid or a thin gold layer in contact with the liquid could be deposited on top of the SiO_2 layer (Figure 16).

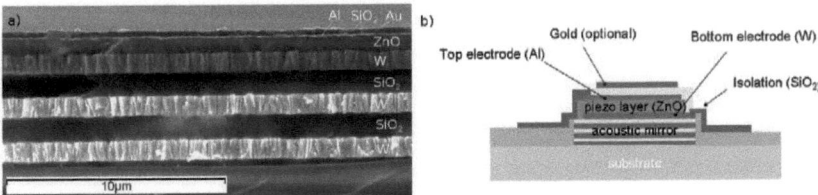

Figure 16: (a) SEM picture and (b) an illustration showing the FBAR stack: The piezoelectric ZnO is sandwiched between an aluminium top electrode and a tungsten bottom electrode. The resonator is built on top of an acoustic mirror consisting of alternating layers of tungsten and SiO_2. The resonator is isolated from the liquid environment with another layer of SiO_2. Some resonators were available with an additional gold layer.

A network analyser (Agilent 8720ES) was used to scan the electrical impedance of the resonators. The FBARs were contacted via microwave probes Model 40A from GGB INDUSTRIES, INC. (Naples, Florida, USA). The serial resonant frequency was read-out by determining the frequency with the maximum electrical conductance. The Q-factor and the resulting dissipation D=1/Q was determined following the formula $Q_s = \dfrac{f_s}{2} \left. \dfrac{<Z}{df} \right|_{f_s}$, where f_s is the serial resonant frequency and $\left. \dfrac{<Z}{df} \right|_{f_s}$ the phase of the electrical impedance at resonance [37, 49]. The resonator characteristics such as electrical impedance, resonant

frequency and mass sensitivity could be simulated using a one-dimensional transmission line model by Mason [124, 125]. Resulting from the set-up described above, the fundamental resonant frequency was about 900 MHz with a simulated mass sensitivity of 1.47 kHz cm^{2}ng^{-1} for the resonators with the SiO$_2$ surface, and about 800 MHz with a simulated mass sensitivity of 1.32 kHz cm^{2}ng^{-1} for the resonators with the additional gold layer. Both types of resonators had a Q-factor of around 160 in water, which was sufficient for measurements in liquid environment. From the resonators covered with gold, either the fundamental frequency or the third overtone at 2 GHz could be read-out. The Q-factor of the third overtone was at around 30 and the simulated mass sensitivity was 2.15 kHz cm^{2}ng^{-1}. Due to the significantly lower Q-factor, the noise of the resonant frequency and particularly the noise of the dissipation signal were much higher than at the fundamental. The FBARs were cleaned in oxygen plasma for 5 minutes at 100 W before the measurements. A flow cell (about 60 µl volume) was mounted on top of the resonators. The set-up is described in detail in [38]. All adsorption measurements were performed at 25° C. All injections were done with a minimum amount of 1 ml on FBAR and 400 µl on QCM-D to ensure complete exchange of the liquids. FBAR measurements were started by recording a baseline for at least 5 minutes followed by injection of the first sample. Measurements were performed at least in triplicate if not stated otherwise.

Figure 17: The FBAR set-up: A syringe is connected to the inlet (1). The FBAR is embedded in one of the PDMS pads (2) and the flow cell (3) with the outlet (4) is mounted on top of it. An arrow shows the position where the RF probe would be connected.

3.3 QCM

A QCM-D (Q-Sense AB, Sweden) equipped with an axial flow chamber was used unless otherwise stated (i.e. Chapter 7.2 and 8 where other types were used). This technique is described in detail in [92]. The change of the resonant frequency Δf and dissipation change ΔD were monitored for several harmonics (5, 15, 25, 35, 45, 55 and 65 MHz) simultaneously. In all diagrams, the normalised frequency shift $\Delta f_n = \frac{\Delta f}{n}$, where n is the overtone number is shown. Prior to their use, the crystals were cleaned for 30 minutes in 2% Sodium Dodecyl Sulphate (SDS) solution and 30 minutes in an UV/Ozone cleaner. All injections were done manually with a syringe with a minimum amount of 400 µl on QCM-D to ensure complete exchange of the liquids. The temperature was stabilised at 25 °C during the measurement.

3.4 SPR

A Biacore 3000 instrument (Biacore AB, Uppsala, Sweden) was used for all SPR measurements shown in this thesis. Thin glass slides were coated with a 50 nm thin gold layer in-house by RF magnetron sputtering, using a two-target sputter coater (Edwards E306A, BOC Edwards, Crawley, West Sussex, UK). A layer of Indium oxide, of approximately 10 nm thick, was sputtered on the glass surface to improve adhesion and chemical resistance of the gold layer to ammonia/peroxide treatment [24]. The gold slides were always cleaned in a boiling solution of hydrogen-peroxide-ammonia in water (1:1:5) and rinsed with water prior to their use in the measurement. The slides were mounted in a plastic chip cassette by double-sided tape and inserted into the Biacore instrument.

3.5 Simulations of the Electrical Impedance Spectrum, Frequency Shift and Mass Sensitivity for QCM and FBAR

Qtools software (Q-Sense AB, Sweden) was used to calculate material properties (e.g. thickness and viscosity) of the adsorbates measured by QCM. Software written in MATLAB (The MathWorks Inc., Natick, Massachusetts, USA) based on the model by Mason [124, 125] was used to simulate the frequency response of both QCM-D and FBAR to adsorbates with different thickness, density, viscosity and elasticity in liquid. The electrical phase versus frequency and thus the resonance frequencies for the FBARs with and without additionally deposited layers with varied thicknesses were simulated using this model. As an input the model needed mass density, acoustic velocity, dielectric constant and intrinsic material Q-factor of each material used in the FBAR stack and the piezoelectric coupling constant of the piezoelectric layer (i.e. ZnO). The output of the simulation was the electrical conductance as a function of the excitation frequency. For details about the model and the exact material parameters used the reader is referred to [49].

The serial resonance frequencies were determined by selecting the frequency at which the electrical conductance reaches a maximum. The model is described in detail in [49].

3.6 Functionalisation for DNA detection

Self-assembled monolayers on the SPR were constructed by manually dispensing the probe and lipoamide in a clean room environment using a pipette. Firstly, a 7 µM thiolated oligo solution was dispensed on a pre-cleaned gold slide, then the probes were allowed to assemble for 15 minutes, the surface was flushed with fresh MilliQ-water. Secondly, a 700 µM solution of Lipa-DEA was dispensed on the surface, incubated for 15 minutes and finally the surface was flushed with water to remove loosely bound molecules. Manually dispensed surfaces were stored at +4°C in humid atmosphere for 3-6 days before SPR measurements due to the fact that storage of the layers has been found to improve the blocking behaviour [25].

The FBARs were functionalised using a piezo dispenser (sciFLEXARRAYER S5, Scienion AG, Germany). Binary solutions of the probes, the DMT-S-S-CALCA-25 and DMT-S-S-PTGS2 at a concentration of 7 µM and Lipa-DEA at a concentration of 700 µM, were dispensed on the cleaned FBAR gold surfaces. Reference pixels were spotted only with 700 µM Lipa-DEA. The spotting procedure was performed at 15°C and 50% humidity. The functionalized chips were kept in the same environment for at least one hour after the spotting in order to give a sufficient time for the probes to bind to the gold surface. The FBARs were rinsed with water, dried and stored for 1 to 3 days at +4°C.

About 1 nl drop volume was sufficient to cover the complete squared gold electrode and not to leave parts of the gold surface uncovered which would significantly increase non-specific binding. The drops also remained separated on the chips so that the different localized functionalisations did not intermix on the chip (Figure 18).

The surface was rinsed with buffer for 10 minutes. If more than one concentration of the complimentary DNA was measured, the surface was regenerated by rinsing the surface for 2 minutes with 10 mM NaOH-0.1% SDS-solution followed by a 2 minute buffer rinse, and the procedure repeated twice.

Human blood serum (Sigma Aldrich) was diluted 1:100 with buffer in order to avoid measurement artefacts caused by changes in viscosity or refractive index.

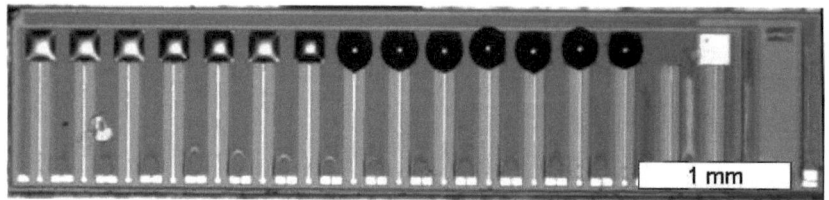

Figure 18: Picture of a 14-pixel FBAR array with droplets of liquid spotted on each pixel. The first 7 pixels from the left

4 The influence of the acoustic properties of the sensor materials on the FBAR performance[2]

While the QCM is widely used as a mass balances for small masses for decades like in biosensor applications [20] or for thin-film deposition monitors [126] the FBAR is novel. The linear relationship between adsorbed mass and resonant frequency is described by the Sauerbrey equation [16] but is only valid if the thickness of the adsorbed film is small compared to the thickness of the resonator and the set-up of the resonator is rather simple (e.g. the electrodes are thin compared to the rest of the resonator). With the Mason model (See Chapter 3.5) the behaviour of acoustic resonators can be simulated also for thick films if the material parameters (i.e. mass density and acoustic velocity) are known. It also allows the simulation of resonators consisting of many layers with comparable layer thicknesses, as is the case for thin-film bulk acoustic resonators (FBAR). The performance of thin-film devices like the FBAR, e.g. the reflectivity of an acoustic mirror, the resonant frequency of a resonator or the mass sensitivity of a mass sensor depends strongly of the acoustic impedance of the materials used. Because the properties of thin-films can significantly deviate from the bulk values, the materials have to be measured as thin-films. For this, a technique exploiting the overtones of thickness-mode resonators has been presented [127]. It is known that the quartz crystal microbalance with dissipation monitoring can be used to determine the viscoelastic properties of polymer thin-films [24, 128].

In this chapter, the influence of the thickness and the viscoelastic properties of the adsorbate on the FBAR frequency response are investigated. For this, a variety of materials are deposited on the resonator in layers of various thicknesses. The influence of the thickness, the acoustic velocity, mass density and the elastic shear modulus to the resonant frequency are determined.

In addition to these materials with known bulk values, the acoustic properties of carbon nanotube thin-films were investigated. Mansfeld et al. [129] used microwave spectroscopy to study the acoustic properties of carbon nanotube films and obtained 2.05 +- 0.05 g/cm^3 for the mass density and between $2*10^4$ m/s and $8*10^3$ m/s for the acoustic velocity,

---

[2] Parts of this chapter will be submitted for publication as Nirschl, M.; Sickert, D.; Schreiter, M.; Vörös, J. Frequency response of thin-film bulk acoustic resonators to the deposition of tungsten, platinum, aluminium oxide and carbon nanotube thin-films. *Accepted for publication in Micro and Nanosystems* **2011**.
The CNTs and the SEM picture were obtained from Oliver Jost (TU Dresden)

depending on the age of the sample, while leaving unclear what exactly happens during this aging process. Weber [130] suggested a very low acoustic velocity of the carbon nanotube layer he deposited on an FBAR operating in the longitudinal mode as an explanation for an increase in resonant frequency after deposition.

An acoustic velocity significantly lower than the one of the solid thin-films (Al_2O_3, W and Pt) was found in this study for the carbon nanotube (CNT) films both for the shear and longitudinal mode.

The method presented in this paper is useful for the measurement of the acoustic properties of thin-films and especially useful for the measurement the acoustic properties of nanostructured thin-films for usage in acoustic thin-film devices. The low acoustic velocity of the CNT films demonstrates the possibility of exploitation of this technique to find micro- or nanostructured materials with acoustic properties significantly different from the commonly used solid thin-films.

4.1 Experimental Section

4.1.1 FBAR

The FBARs used for the measurements in this chapter had a less complex set-up to the ones described in the experimental section in Chapter 3.2. The ZnO layer is sandwiched between the tungsten bottom electrode and a gold top electrode. The top electrode was not isolated from the environment, which makes it impossible to use this design in liquid, which was not necessary in this chapter. Like all resonators used in this thesis they were mounted on top of an acoustic mirror, electrically excited through the electrodes and vibrated in shear mode. The set-up and a SEM picture of one of the resonators are shown in Figure 19.

The resonance frequencies were read out using a Network Analyser as described in Chapter 3.2.

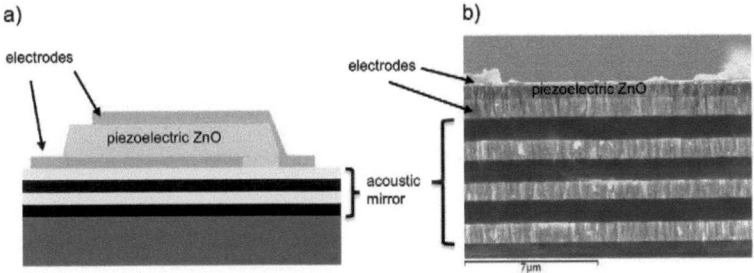

Figure 19: a) Sketch and b) scanning electron microscope picture of the FBAR used in Chapter 4: A piezoelectric ZnO layer is sandwiched between two electrodes and mounted on top of an acoustic mirror.

4.1.2 Thin-film Deposition

Al_2O_3, W and Pt thin films were produced by magnetron sputtering (Perkin Elmer). Al_2O_3 films were deposited at 100 W and 5 mTorr from an Al_2O_3 target. Process time was varied from 10 min to 160 min, which resulted in film thicknesses from 20 nm to 375 nm. Pt was sputtered at 100 W and 13 mTorr with process times from 1 to 20 min resulting in thicknesses from 30 to 225 nm. W films were deposited at 25 mTorr and 600W with process time varied from 1 min to 10 min resulting in thicknesses from 30 nm to 300 nm. The Pt and Al_2O_3 layers were structured using a standard lift-off process, the tungsten layers by a physical back etching process. The thickness was measured with a profilometer (Dektak) at least three times at different positions. A layer was deposited and the resonant frequency shift was measured before and after the deposition on at least three different resonators for each thickness and material.

4.1.3 Carbon Nanotube Films

Purified single-walled carbon nanotubes were obtained from Institute of Material Science, TU Dresden. The carbon nanotubes were produced by LASER evaporation and purified by an acid wash, followed by filtration and resuspension in deionised water [131, 132]. The carbon nanotube films were produced similar to those described in [133]: The carbon nanotubes were dissolved in natural bile salt detergent deoxycholic acid (DOC) [134] under ultrasonication and centrifuged for 30 min at 12000 rpm. The supernatant was recovered and repeatedly centrifuged for three times to remove excess unsuspended carbon nanotubes. The suspension was then vacuum-filtered onto a filtration membrane. The nanotube carrying membrane was pressed face-down onto the substrate, i.e. the resonator surface, and dried at 105 °C for 60 min. Afterwards the substrate was immersed in acetone to dissolve the membrane and leave the nanotube film adhered to the surface.

In order to avoid short-circuiting the top and bottom electrode and to provide structures for measuring the film thickness with a profilometer, the carbon nanotube films were structured by covering the resonator area with photo resist and removing the uncovered nanotubes by exposing them to oxygen plasma.

4.2 Simulation of the Frequency Response for Thin-film Adsorption of Materials with Different Acoustic Velocities and Mass Densities

The electrical phase of the FBAR with material properties as described in the experimental section in Chapter 3.2 was simulated from 600 MHz to 2.2 GHz (Figure 20 a). The simulation shows four shear resonances (black curve) and one longitudinal resonance (grey line) in this frequency range. Figure 20 b) shows the measured electrical phase from 600 MHz to 2.2 GHz in air and water. The five resonances found in the simulations can also be seen in the measurement in air (grey curve), where two resonances at around 1.8 GHz partly overlap. One of the resonances is nearly fully damped when the sensor is put into contact with water (black line) indicating that this is the longitudinal mode. The other resonances are reduced less, showing that these are shear mode resonances. The position of the resonance frequencies of both the longitudinal and the shear modes are consistent with their positions in the simulated curve.

Unlike the quartz crystal microbalance (QCM), where the overtones f_n are equidistant at $f_n = f_0 * n$, with n being the overtone number where only odd numbers are permitted, the FBAR has resonances at less regular positions due to the complex layer structure. The many layers forming the FBAR provide numerous interfaces between layers with different acoustic impedance where acoustic waves are reflected and resonances with irregular frequency spacing can occur.

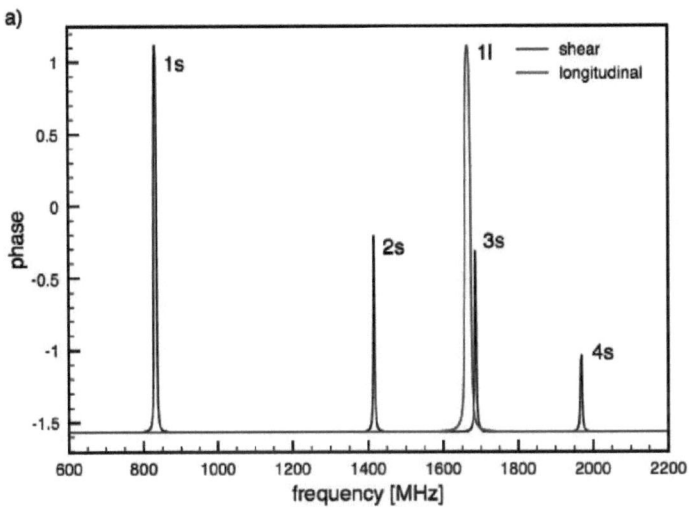

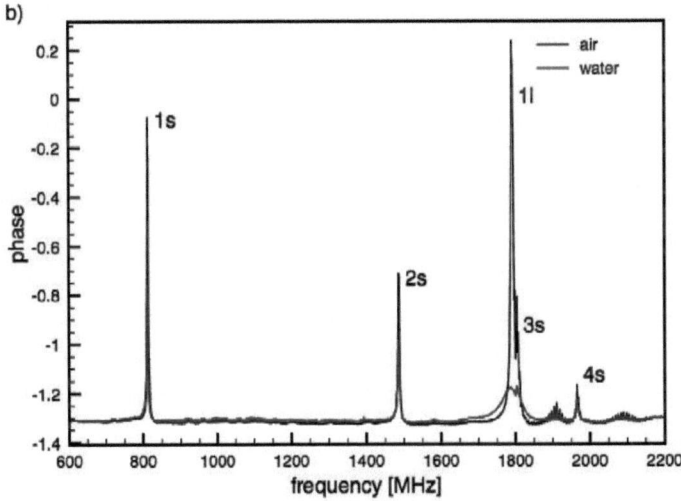

Figure 20: Frequency spectrum of the FBAR: a) simulated shear and longitudinal modes and b) measured in air and water. Up to five shear modes (1s, 2s, 3s, 4s and 5s) and one longitudinal mode (1l and 2l) were monitored in this study. 5s and 2l are not shown in this figure, as they were not visible with the FBAR used to record this frequency spectrum.

The changes in resonant frequency caused by the deposition of layers with various densities and acoustic velocities were simulated.

Figure 21 a) shows the frequency behaviour upon adsorption of materials with different densities (2000 to 20000 kg/m^3) but equal acoustic velocity (2800 m/s). The density range includes all densities of the materials used in this paper for both the FBAR fabrication and the added layers. The selected acoustic velocity is approximately that of the piezoelectric material in the FBAR (ZnO). The frequency response to the adsorption of the additional layer is different for the different densities. Layers with the same thickness but higher density cause a higher frequency shift which can be seen by a higher inclination of the curves in the diagram, as predicted by the Sauerbrey equation [16]. Figure 21 b) shows the simulated adsorption of two materials with fixed densities of 2000 kg/m^3 and 20000 mg/m^3, respectively, but with acoustic velocities varied between 400 m/s and 2200 m/s. For small thicknesses of up to around 50 nm the acoustic velocity does not influence the frequency shift and the frequency shift changes linearly with thickness depending on the density. At thicknesses higher than 50 nm, a deviation from the linear behaviour can be seen: For 2000 kg/m^3, the lower the acoustic velocity, the higher the deviation from the linear behaviour towards larger frequency shifts. For 20000 kg/m^3, the higher the acoustic velocity, the higher the deviation from linear behaviour towards lower frequency shifts.

An analysis of the part of the acoustic wave inside the resonator and the part contained in the additional layer might help to explain this behaviour (Figure 22). Because there is only a small part of the acoustic wave inside the added films when they are very thin, the behaviour of the resonator is mainly influenced by the acoustic properties of the materials in the resonator (Figure 22 a). With increasing thicknesses of the added layer, a larger fraction of the acoustic wave is inside the added layer and not predominantly inside the resonator (Figure 22 b).

This happens more readily for materials with lower acoustic velocity (Figure 22 c).

If a significant part of the acoustic wave is inside the added layer, the acoustic properties of the added layer significantly influence the resonator behaviour.

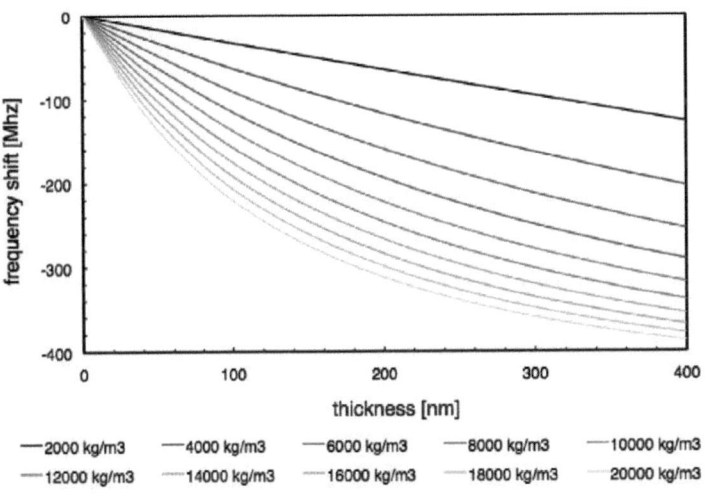

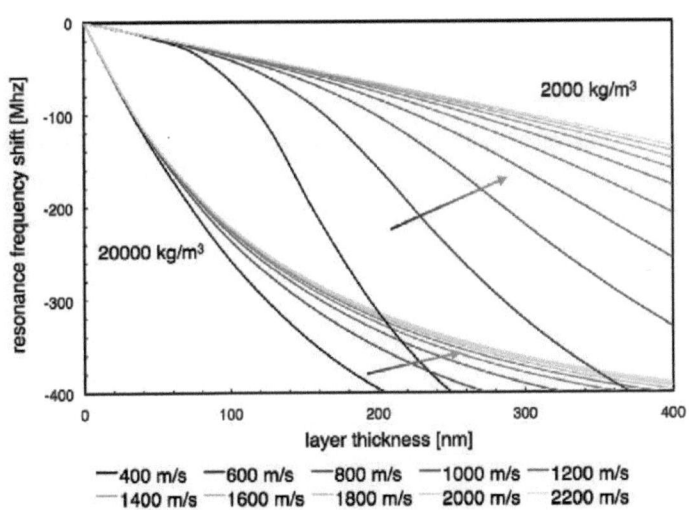

Figure 21: Simulation of the adsorption of thin-films on top of the resonator with a) acoustic velocity of 2800 m/s and mass densities varying from 2000 kg/m^3 to 20000 kg/m^3 and b) equal mass density of 2000 kg/m^3 and 20000 kg/m^3, respectively, and acoustic velocities varied from 400 m/s to 2200 m/s. The arrows point towards higher acoustic velocities.

The acoustic impedance Z_{ac} is given by $Z_{ac} = \rho_i v_i$, where ρ_i is the density and the v_i acoustic velocity of the added layer material. With the sensitivity S_{res} being $S_{res} = \left|\frac{\Delta f}{\Delta m}\right| = \frac{2 f_0^2}{A} \frac{1}{Z_{ac,res}}$, where Δf is the frequency shift caused by adding a rigid mass Δm, $Z_{ac,res}$ the effective acoustic impedance of the resonator, f_0 the fundamental resonant frequency of the resonator and A the area of the resonator, a lower acoustic impedance and thus a lower acoustic velocity increases the mass sensitivity of the resonator [135]. A material with a high acoustic velocity instead decreases the mass sensitivity.

In this way, if a material with a high acoustic velocity is added, the effective acoustic impedance is increased, the mass sensitivity is decreased and further added mass causes a smaller frequency shift as on the uncovered resonator. If the acoustic velocity of the added layer is lower than the one of the resonator, the opposite happens: The effective acoustic impedance of the resonator decreases and further added mass results in a higher frequency shift than on the uncovered resonator: the mass sensitivity of the resonator is increased.

As a result, an additional layer with acoustic impedance as small as possible should be added to maximise the mass sensitivity of the resonator. The higher the thickness of this added layer, the more the acoustic impedance of the overall stack comprised of resonator and added layer differs from the resonators impedance and approaches the impedance of the added layer. That means that the highest theoretically reachable mass sensitivity $S_{res,max}$ is $S_{res,max} = \frac{2 f_0^2}{A} \frac{1}{Z_{ac,adlayer}}$, where $Z_{ac,adlayer}$ is the acoustic impedance of the added layer. As a simple example, if the acoustic impedance of the added layer is 10 times smaller than the one of the resonator, the sensitivity might increase by a factor up to 10. However, the limit of detection, that depends both on the mass sensitivity and the noise, is more appropriate for evaluation of the sensor's performance [136]. The frequency noise might increase e.g. due to intrinsic energy dissipation in the added layer or a reduced piezoelectric coupling coefficient and thus the limit of detection might not improve as much as the mass sensitivity.

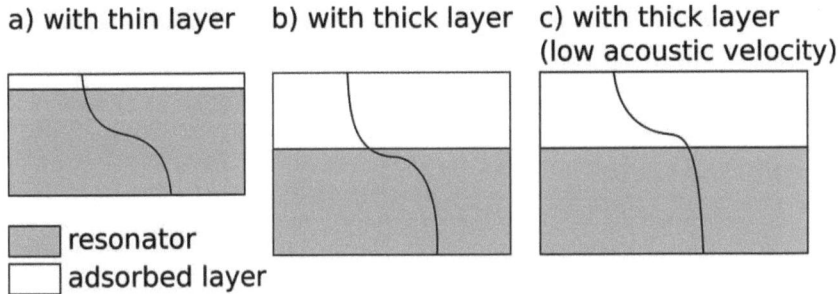

Figure 22: Simplified sketch of the acoustic wave in the resonator and a) a thin added layer, b) a thick added layer with acoustic properties close to the ones of the resonator and c) a thick added layer with an acoustic velocity lower than the one of the resonator.

4.3 Deposition of Platinum, Aluminium Oxide, and Tungsten Thin-films

In order to investigate if the influence of thickness, density, and acoustic velocity on the resonant frequency observed experimentally agrees with that predicted in the simulations, tungsten, platinum, and aluminium oxide thin-films were deposited on the resonators with thicknesses ranging from few tens to few hundreds of nanometres. The measured shift in resonant frequency as a function of the film thickness can be seen in Figure 23 (symbols). The dashed lines show the curves that were fitted to the measured values. The fitted values for density and acoustic velocities are shown in Table 4 together with values from the literature.

The measured mass density of Pt corresponds well to the bulk values from literature and is consistent with previous measurements on thin-films [137]. The values for the density of W and Al_2O_3 are slightly lower than the bulk values from literature [49], which result from earlier measurements on thin-films e.g. by Rutherford Backscattering Spectrometer or by weighing the thin-films on a balance [138, 139]. As shown in Figure 21 b), the difference in the frequency shifts between layers with different acoustic velocities is smaller for higher acoustic velocities, especially for higher densities, because the largest part of the acoustic wave remains in the resonator. With the given measurement inaccuracy, it was not possible to distinguish between the fit using different acoustic velocities and the measured values for the W, Pt and Al$_2$O$_3$ films. Only a lower limit for the acoustic velocities could be estimated.

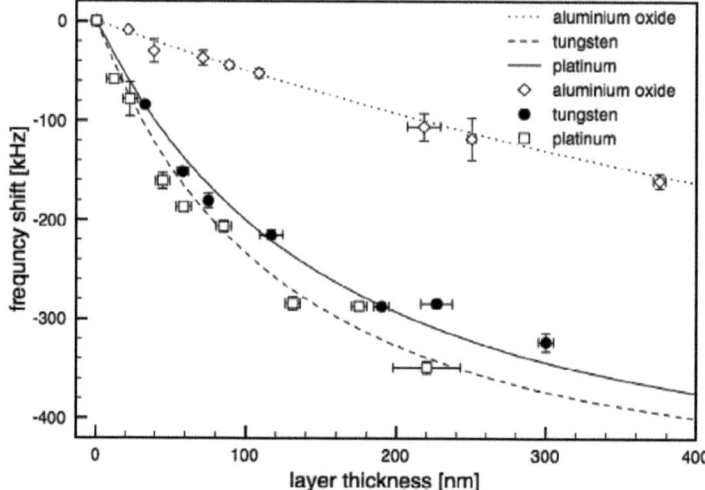

Figure 23: Measured values (symbols) of the resonant frequency shift on adsorption of three different materials (Al$_2$O$_3$, Pt, W) and curves fitted to the Mason model (lines).

	Mass density [10^3 kg/m^3]		Acoustic velocity [m/s]	
	Measured	Literature	Measured	Literature
Al$_2$O$_3$	3.1 ± 0.1	3.9	> 2000	5790
Pt	22.0 ± 0.7	21.5	> 800	1690
W	17.2 ± 0.3	19.3	> 1000	2840
CNT	1.5 ± 0.1	2.05 ± 0.05 [129]	653 ± 13 (shear) / 810 ± 36 (longitudinal)	N/A (shear) / 80 (longitudinal)

Table 4: Summary of the values for mass density and acoustic velocity of aluminium oxide, platinum, tungsten (all bulk values) and carbon nanotube thin-films. For the CNT thin-films, a value is shown for each the shear and the longitudinal mode.

4.4 Deposition of Carbon Nanotube Films

Analogous to the measurement with the solid thin-films, CNT films with thicknesses from about 10 to 250 nm were deposited on the FBARs.

Figure 24 a) shows a SEM picture of the CNT film. The amount of suspension filtered onto the membrane was varied from 20 ml to 140 ml resulting in 7 different thicknesses up to 250 nm (Figure 24 b).

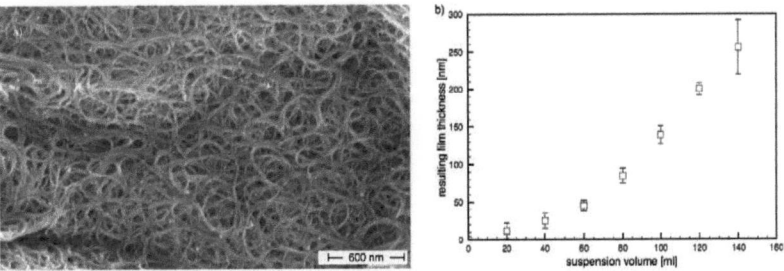

Figure 24: a) Scanning electron microscope picture of a carbon nanotube film. b) The thickness of the CNT layers versus the amount of CNT solution used.

Figure 25 shows the measured frequency shifts of all shear (black symbols) and longitudinal modes (red symbols) in the frequency range from 500 MHz to 2.7 GHz together with curves fitted by least squares regression to the measured values.

All shear and longitudinal modes were fitted simultaneously. The values found from the fit are $1.5 * 10^3$ kg/m^3 for the mass density, 653 m/s for the acoustic shear velocity and 810 m/s for the acoustic longitudinal velocity.

The fitted curves correspond well to the measured values. The frequency does not decrease linearly; instead, the inclination decreases with increasing thickness. At about 200 nm layer thickness, two resonances can be seen within the frequency range from 400 to 1400 MHz. One of them is the fundamental shear mode (1s) and the other one the next higher shear mode (2s) whose resonant frequency is greatly decreased. For a thickness of about 250 nm, the height of the peak of the fundamental resonant frequency is too small to be detected and only the higher mode can be seen.

The frequency decreases until a thickness is reached where a quarter of the wavelength of the acoustic wave is inside the CNT layer. From this point on, the gradient of the curve decreases until the resonant frequency has reached the frequency of the preceding mode without the CNT layer. This occurs for both the shear and the longitudinal modes.

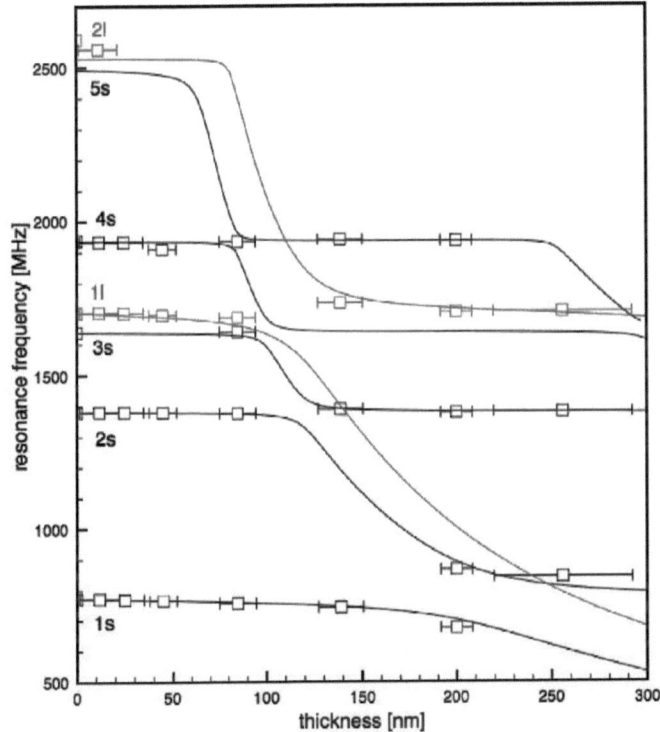

Figure 25: Resonant frequency of all measured shear (1s, 2s, 3s, 4s and 5s) and longitudinal (1l and 2l) resonances between 500 MHz and 3 GHz for different thicknesses of added CNT films. Symbols show the measured values, lines the fitting results. Shear modes are in black, longitudinal modes in red.

4.5 Enhancement of FBAR Mass Sensitivity using Materials with Low Acoustic Impedance such as CNTs

Previous simulations and measurements in this section suggest that upon adding a material with low acoustic impedance the mass sensitivity can be significantly increased. The simlulation was performed as in the previous chapter (Chapter number), the sensitivity is derived from the slope of the adsorption curves. Figure 26 shows the simulated mass sensitivity of the FBAR if a layer of CNT is added on top. All modes show an increasing sensitivity with increasing layer thickness until the sensitivity of all modes reaches a maximum. The maximum mass sensitivity for the first shear mode is 73 kHz*cm^2/ng, more than one order of magnitude greater than the sensitivity of the resonator without CNT (4.4 kHz*cm^2/ng). The largest mass sensitivity (270 kHz*cm^2/ng) of all modes is reached for the second longitudinal mode with a 90 nm thick CNT layer. These simulations, however, do not provide any information about the Q-factors of the resonators with the CNT layers. It is likely that the Q-factor decreases with increasing CNT thickness, and it might even be that at the theoretical maximum of the sensitivity the Q-factor is too small, and therefore cannot be read-out anymore. In other words, if one seeks to increase the limit of detection of the FBAR by adding CNTs, a compromise between high mass sensitivity and low Q-factor (and thus higher noise) must be found.

An additional increase in sensitivity might arise from the fact that the CNTs have a higher surface roughness than e.g. gold and therefore a higher effective surface area.

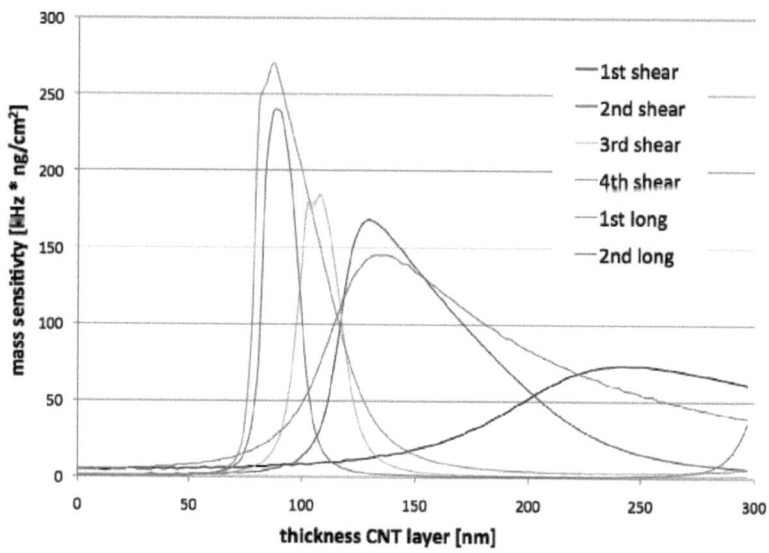

Figure 26: Mass sensitivity changes for different modes by adding CNT layers with thicknesses up to 300 nm.

4.6 Summary and Outlook

The simulations show that materials with different acoustic properties (density and acoustic velocity) cause the FBAR to respond differently. While the response is linear and depends on the layer density for very thin films (for the materials used in this study lower than 50 nm), there are deviations for thicker films depending on the acoustic velocity of the added material. In this way, thin films with different mass density or different acoustic velocities can be distinguished by the resonant frequency shift they induce. If the deposition of a film is continuously monitored (e.g. adsorption of material) this could provide extra information given a known film thickness.

The frequency shift caused by deposition of tungsten, aluminium oxide and platinum films on the resonators is clearly different for the three materials. The fitted values for their mass density were in good agreement with the literature. Their acoustic velocities were too high to be distinguished and only a lower limit could be determined. However, in case of the carbon nanotube thin-films, not only the mass density but also the acoustic velocity could be fitted. The fitted values for the mass density are lower than the values from literature, which could be caused by the production process. The acoustic velocities obtained for both the shear and the longitudinal mode are far lower than for the other materials used in this study. This makes CNT thin-films a promising material for devices in which materials with low acoustic impedance are desired such as acoustic Bragg mirrors or mass sensors.

The method presented is especially interesting if the materials under investigation are to be potentially used in the FBAR because in this case the material parameters are determined with exactly the same device in which they will be used. In this way, new materials with unknown acoustic properties, and their suitability for the FBAR can be evaluated. These materials could be, for example, materials with a micro- or nanostructure with acoustic properties that differ from to the unstructured bulk material.

While the measurements in this chapter were performed in air, the results for the CNT and the simulations of their use in for making the device more sensitive are also valid for devices operating in liquid. In this way the use of micro-structured materials might be a way to decrease the sensitivity gap between the FBAR and the most sensitive transducers (See Chapter 1.1 and Table 1).

5 Measurement of DNA and Protein Adsorption on Passive FBAR[3]

In this chapter, the first measurements that were performed with the FBAR in order to test the whole set-up and evaluate the performance of the technology are described. Firstly, the mass sensitivity of the passive FBAR was evaluated in a comparison with the SPR. The FBAR was used for the detection of proteins and DNA and the kinetics of adsorption and recrystallisation of bacterial surface layer protein (S layers) monomers into two-dimensional (2D) protein crystals at gold surfaces were investigated.

5.1 Experimental Section

5.1.1 S Layer Proteins

S layers form the outermost cell envelope component of many bacteria and archaea. They can exhibit different lattice symmetries with lattice periodicities around 10 nm. S layers are typically 5 - 10 nm thick having identical, regularly arranged pores with diameters of 2 - 8 nm. The possibility of reconstituting isolated S-layer subunits in vitro into 2D arrays with perfect uniformity over large areas on solid surfaces or at liquid-air interfaces [140, 141] makes them an almost ideal biological template for supra-molecular engineering. Due to their periodic structure, they constitute an ideal template for the bottom-up fabrication of periodic cluster arrays with very high density [142-150]. Amongst others, these cluster arrays are well suited for catalytic sensor applications [151, 152].

The S layer was isolated from the bacterium Bacillus sphaericus NCTC 9602. The conditions for cell cultivation and purification of S-layer sheets were described previously [149]. In order to disintegrate the protein sheets into monomers, 500 µl of the S-layer

---

[3] Parts of this chapter were published in Nirschl, M.; Blüher, A.; Erler, C.; Katzschner, B.; Vikholm-Lundin, I.; Auer, S.; Vörös, J.; Pompe, W.; Schreiter, M.; Mertig, M. Film bulk acoustic resonators for DNA and protein detection and investigation of in vitro bacterial S-layer formation. *Sensors and Actuators A: Physical* **2009**, *156*, 180-184 or Mertig, M.; Bluher, A.; Erler, C.; Katzschner, B.; Pompe, W.; Nirschl, M.; Schreiter, M. Investigation of in-vitro bacterial surface layer formation by FBARs. *2009 IEEE Sensors* **2009**, 1161-1164 or Auer, S.; Nirschl, M.; Schreiter, M.; Vikholm-Lundin, I., Detection of DNA hybridisation in a diluted serum matrix by surface plasmon resonance and film bulk acoustic resonators. *Analytical and Bioanalytical Chemistry* **2011**, 1-10.
S-Layer proteins were obtained by Anja Blüher (TU Dresden); DNA surface chemistry and SPR measurements were provided by Inger Vikholm-Lundin and Sanna Auer (VTT Finnland)

suspension was washed twice with distilled water and centrifuged for 45 min at 25.000 g and 4°C. Then the pellet was incubated in 300 µl of 6 M guanidinium hydrochloride for 2 h at room temperature. Thereafter, the denatured protein suspension was twice dialysed against TRIS/HCl buffer, pH 9.0, for 45 min. This solution can be stored at 4°C for at least 7 days. To initialize recrystallisation at substrate surfaces, MgCl2 was added with a final concentration of 10 mM.

5.2 FBAR Mass Sensitivity Comparison with SPR

In order to determine the mass sensitivity of the applied FBAR in ng/cm^2, the surface coverage of the sensor pixel was calibrated against a well-established technology with known mass sensitivity, the SPR. To obtain the mass sensitivity of the FBAR, both sensor systems were exposed to bovine serum albumin (BSA) solutions of increasing concentration. The serial resonant frequency shift, f_s, of the FBAR and the change in reflective angle of the SPR device were simultaneously monitored. The result of this measurement is plotted in Figure 27. By comparing the FBAR frequency shift to the surface coverage obtained by SPR at the corresponding BSA concentration, a mass sensitivity of ~2 kHz cm^2/ng was estimated. The minimum detectable mass, which corresponds to the threefold standard deviation of the baseline noise, divided by the sensitivity was found to be in the range of ~1 ng/cm^2, about two orders of magnitude higher than for the SPR.

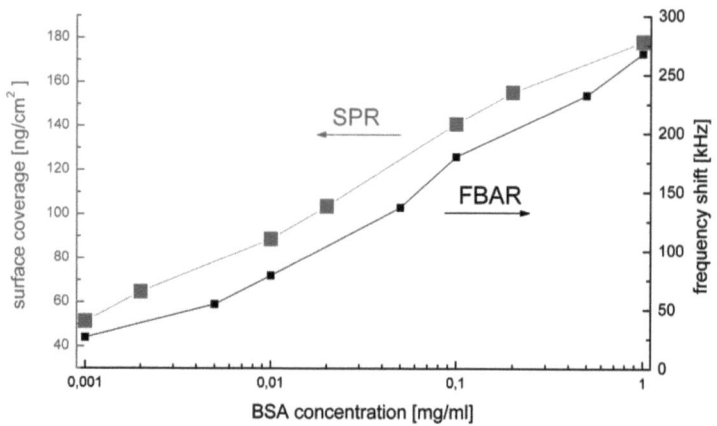

Figure 27: BSA adsorption on gold as measured by surface plasmon resonance and the FBAR sensor chip. The mass coverage obtained with the SPR device is plotted together with the detected resonant frequency shift of the FBAR.

5.3 Measurement of Protein-Protein Interaction using a Dual Measurement Probe

As one of the interesting features of the FBAR is the possibility of integration of many FBAR sensors on a small area in one sensor device, simultaneous measurement of the response of two FBARs at was demonstrated. In this the set-up a dual measurement probe was used (see Figure 28), similar to the one described Chapter 3.2 rather than a single probe and simultaneous real-time measurement of two FBARs in liquid were performed.

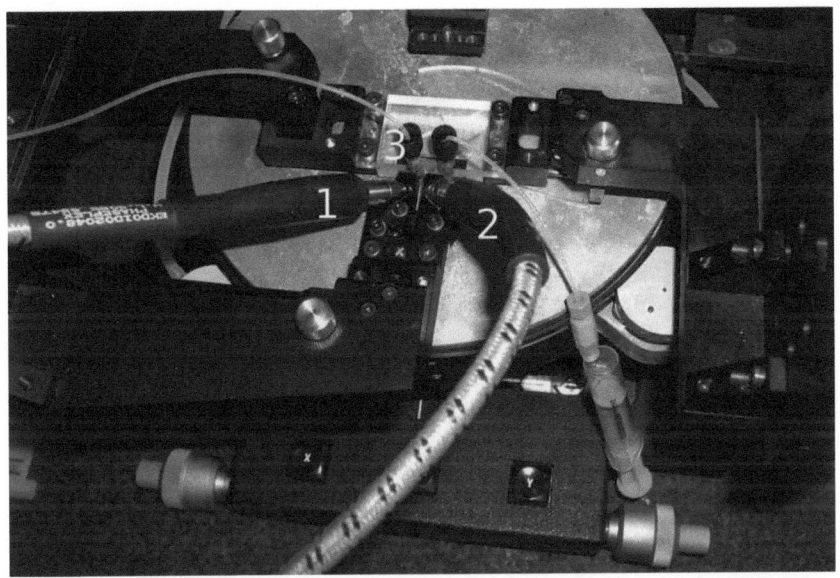

Figure 28: The dual probe set-up: Instead of one picoprobe, a dual probe (1 and 2) is connected to the FBAR under the flow cell (3).

As an example measurement one pixel was functionalised with mouse-IgG and a second one with goat anti-IgG.

Figure 29 shows the measurement of the two pixels (Both of them 200 µm x 200 µm large): First the remaining gold area that was not covered by anti-IgG was blocked with 0.5 mg/mg BSA. Then both sensors were rinsed with mouse IgG, which only bound to the pixel functionalised with mouse anti-IgG and the one functionalised with goat anti-IgG remained stable. Finally, both sensors were rinsed with goat IgG, which only bound to the pixel functionalised with goat anti-IgG and no binding was observed from the one functionalised with mouse anti-IgG. This first measurement using two FBARs at the same time demonstrated the ability of the FBAR technology to be used in a multiplexed format. Later measurement using 64 pixels at the same time using CMOS technology are described in chapter 7.

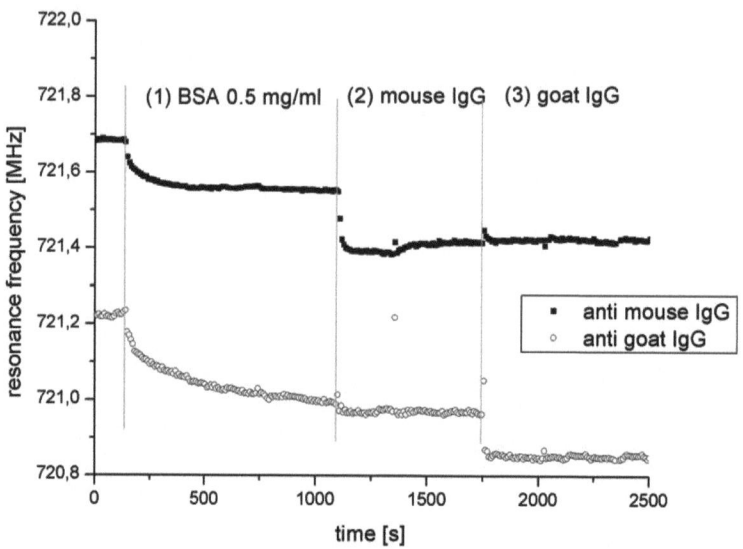

Figure 29: Parallel monitoring of two FBAR pixels – one with immobilised mouse anti-IgG, the second with immobilised goat anti-IgG. The sensors are rinsed with BSA first, then with mouse IgG and finally with goat IgG. It can be seen that BSA adsorbs on both pixels (1), mouse IgG binds only at the sensor functionalised with mouse anti-IgG (2) and the goat IgG binds only on the pixels functionalised with goat anti-IgG (3). Buffer rinsing steps are not shown for clarity.

5.4 DNA Detection

In addition to the protein detection described in Chapter 5.3 this chapter also details the measurement of DNA hybridization, which is interesting and critical in several areas of analysis: molecular biology, genotyping, diagnostics and screening of genetically modified food ingredients [153, 154]. In this chapter we compare the FBAR performance to the SPR, which has previously been used for real-time detection of DNA hybridization in the nanomolar range[55, 155]. The detection of oligonucleotide interactions has also been measured by QCM but is rather novel on the FBAR [20, 38]. The surface functionalisations for DNA detection are critical but many solutions were published. Based on the pioneering work of Herne and Tarlov [156] on SPR characterisation of thiol-modified ssDNA (single-stranded DNA) probes and mercaptohexanol self-assembled monolayers (SAM) much was learned about the surface functionalisations: The hybridization properties of the target are affected by surface probe length and also by factors such as probe density, surface orientation and target sequence [157-159].

For the measurements shown in this chapter, a surface chemistry which was developed and provided to me by VTT Finland [122, 160]. This chemistry allows the assembling of the probe molecules on the surface *in situ* during 10-15 minutes with good surface coverage of the probes [122, 161, 162].

In order to reduce the non-specific binding, in previous publications thiolated hydrophilic organic molecules like mercaptohexanol and polyethylene-glycol-derived compounds were embedded between the oligos on the surface [156, 159]. These compounds render the surface non-fouling, reduce the non-specific binding and induce the DNA-oligos to maintain an upright orientation on the surface maximising their hybridization capacity [122, 156, 159, 161, 162].

Nonspecific binding has shown to be particuarly problematic when the DNA is being detected from serum [163]. Synthetic DNA ~100 bases long breast-cancer specific DNA strands are used as target strands mimicking single stranded PCR amplicons (ssPCR), and instead of biopsy samples from the actual breast tissue the samples were diluted in a more easily available serum matrix (1:100). The disulphide- modified S-S-DNA probes were attached to the surface from a binary solution with the lipoamide (Lipa-DEA) as a blocking agent.

Binding curves for the same samples of spiked serum measured by both SPR and the FBAR are shown in Figure 30 c. The responses caused by serum (200 ± 50 RU with SPR and 160 ± 20 kHz with FBAR) were subtracted from the values of the DNA spiked samples. The lowest PTGS2-123 ssDNA concentrations (0.01 – 0.1 nM) did not differ much from that of serum. A 1 nM PTGS2-123 ssDNA concentration gave a response of 80 ± 10 RU and 75 kHz as measured with SPR and the FBAR. In the case of CALCA-92 the corresponding responses after the serum baseline subtraction were 60 ± 10 RU and 60 ± 10 kHz, respectively at a 1 nM CALCA-92 concentration (Figure 4d). The SPR response in serum was thus about half of that in buffer at low concentrations.

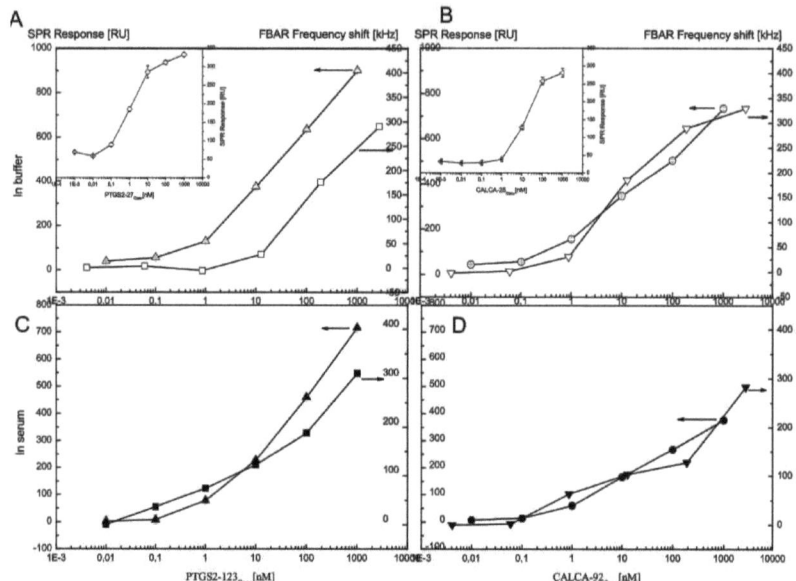

Figure 30: Hybridisation of long ssDNA strands to layers of A, C) DMT-S-S- PTGS2-27/Lipa-DEA or B, D) DMT-S-S-CALCA-25 as measured with SPR (Δ,○) and the FBAR resonator (□,▽) in buffer and serum (open and filled symbols), respectively. The insets show SPR curves for hybridisation with complimentary A) PTGS2-27 and B) CALCA-25 in buffer

The limit of detection (LOD) seems thus to be in the range of 1 nM (calculated as three times the standard deviation/noise) for both FBAR and SPR for samples measured in buffer as well as for samples diluted in serum. Thus long ssDNA strands spiked in serum down to a concentration of 1 nM can be detected. With the surface functionalisation from VTT, the response caused by the hybridisation can be distinguished even for the samples spiked in serum. Comparison of responses and frequency shifts acquired with different samples and sample matrices is presented in Table 5.

Hybridisation with long stranded PTGS2-123 and CALCA-92 PCR products in buffer with the surface layer showed a response of 900 and 740 ± 10 RU at a concentration of 1000 nM. This corresponds to 2.4 and 2.2 ± 0.01 x 10^{12} targets/cm^2 (Table 5 and Table 6). In serum the responses were 20 and 50 % lower than that in buffer for PTGS2-123 and

CALCA-92, respectively. A response of 720 ± 10 RU or 1.4 ± 0.01 x 10^{12} targets/cm^2 was obtained for hybridisation of PTGS2-123 and the response obtained for CALCA-92 was 380 ± 10 RU or 1.5 ± 0.01 x 10^{12} strands/cm^2. The hybridised target densities for PTGS-123 and CALCA-92 were the same in both buffer and serum and similar to those reported by Gong et al. [164] (2.2 ± 0.8 x 10^{12} molecules/cm^2).

When studying only the FBAR responses (Figure 30 a – d), it can be observed that frequency shifts in serum were not only observed at high concentrations like in measurements performed in buffer, but at concentrations as low as 0.01 nM. This indicates that the frequency shifts with the FBAR included non-specific binding of substances from serum. The origin of the non-specific binding might be incomplete passivation of the gold surface. A dispensing procedure used for the functionalization of the FBARs deposits small liquid droplets onto square gold electrodes. This process might leave parts of the electrode uncovered - especially the corners might be prone to incomplete coverge of dispensed liquid. The empty gold surfaces then provide a mass-sensitive area for non-specific binding. To avoid this artefact, the dispensing procedure should be optimized (e.g. increase the amount of liquid dispensed) so that the gold electrodes are fully covered by the dispensed liquid. Further differences between the shape of the SPR and the FBAR curves might come from the different flow cells (Size, flow velocity and cell height) used for the two technologies.

The FBAR responses were the same in buffer and serum for both of the ss-DNA products. A lower response was observed for PTGS2-123 than for CALCA-92 contrary to what was observed with SPR.

	SPR [RU]	FBAR [kHz]	SPR [RU]	FBAR [kHz]
	PTGS2		CALCA	
Conc [nM]	In buffer			
BSA	20 ± 5	< 10	60 ± 20	< 10
Non-comp DNA	30 ± 5	< 10	140 ± 15	< 10
1 nM	130 ± 5	30	350 ± 10 (10nM)	30 ± 20
1000 nM	900 ± 10	120 ± 5	740 ± 10	340 ± 10
	In serum			
1 nM	80 ± 10	50	170 ± 5 (10nM)	35 ± 15
1000 nM	720 ± 10	130 ± 20	380 ± 10	360 ± 20

Table 5: SPR and FBAR responses for hybridisation of short and long DNA strands with probe/Lipa-DEA layers with samples diluted either in buffer or in a serum matrix. FBAR standard deviation was approximately 5-20 % in cells where not shown.

The density of targets on the spotted FBAR was calculated to be similar for CALCA (1.6 ± 0.01 x 10^{12}), but much lower for PTGS2 (0.4 ± 0.01 x 10^{12}). The FBAR results might be different from the SPR because the FBAR is sensitive to the water content [165] and the viscoelastic properties [166] of the DNA layer. The difference between the PTGS2 and CALCA on FBAR can be explained by differing liquid behaviours during the spotting procedure (See e.g. Figure 18), due to differences in the liquid properties and binding behaviour. Investigating this in detail, however, would have been beyond the scope of this chapter, which was to evaluate the FBAR performance with a working DNA functionalisation process in comparison to other sensor platforms (i.e. SPR).

Target density	PTGS2 [molecules/cm^2]	CALCA [molecules/cm^2]
By SPR	$1.4 \pm 0.01 \times 10^{12}$	$1.5 \pm 0.01 \times 10^{12}$
By FBAR	$0.4 \pm 0.01 \times 10^{12}$	$1.6 \pm 0.01 \times 10^{12}$

Table 6: Target densities for hybridisation of long DNA strands at a concentration of 1000 nM to PTGS2 and CALCA probe/Lipa-DEA surfaces calculated by SPR responses or FBAR frequency shifts.

5.5 Adsorption an Recrystallisation of S-Layer Proteins on Gold

In this chapter, the FBAR is used to measure the process of self-assembly of S-layer protein monomers into 2D protein crystals on gold surfaces. Because the FBAR measures molecules adsorbed to the surface without the need of a label or a reference channel, it is easy to use for following adsorption processes to e.g. a gold surface.

For simplicity and to be able to mix several liquids in top of the FBAR, no flow cell was used for this measurement; instead a cell that was open on the top was mounted on the FBAR. With this setup it was possible to pipette small amounts of solution directly to the sensor. The measurement was started after 10 µl of TRIS buffer and 10 µl of 100 mM $MgCl_2$ was pipetted to the sensor. After a stable baseline was reached, 80 µl of S-layer monomer solution was added in concentrations such that the final protein concentrations of the solution varied from 0.05 to 10 mg/ml. The cell was then closed with a lid to avoid evaporation of the solution.

Immediately after the S-layer monomer solution was added to the cell on top of the FBAR, a decrease in the resonant frequency was observed. Figure 31 depicts the initial time dependencies of the observed frequency shifts recorded for the film formation at five different monomer concentrations. The monomer solution was added at t = 0. As expected, the frequency shift measured within the first 5 minutes after the monomer solution was added increases with increasing monomer concentration. The reaction kinetics are accelerated with increasing protein concentration. Figure 32 shows concentration dependence of the initial rate of the monomer adsorption on the gold surface, calculated as the derivative of the frequency shift per time interval at t = 0.

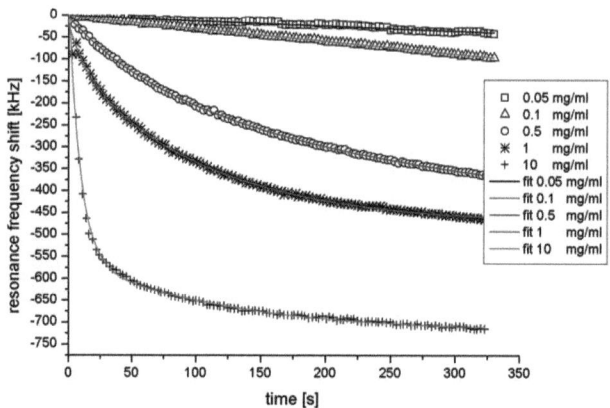

Figure 31: FBAR frequency shifts during the first 325s of adsorption and recrystallisation of S-layer monomers of Bacillus sphaericus NCTC 9602 into 2D protein layers at the gold surface of a FBAR at concentrations ranging from 0.05 to 10 mg/ml. The different symbols are measured data; the solid lines are curves obtained by fitting the experimental data with the function given in Equation 1.

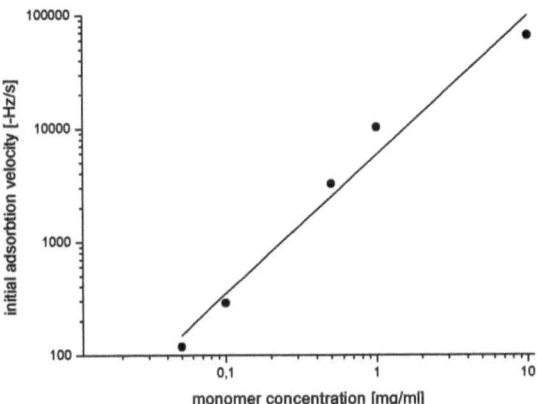

Figure 32: Dependence of the S-layer monomer adsorption velocity at t = 0 on the initial monomer concentration. The linear regression line with a slope of approximately one implies a power-low dependence.

A more rigorous quantitative analysis of the measured frequency shifts revealed that the initial time dependence of the S-layer adsorption and recrystallisation could be fitted with a linear combination of at least two exponential functions:

$$fs = A \cdot \left(-1 + e^{-B \cdot t}\right) + C \cdot \left(-1 + e^{-D \cdot t}\right),$$

where the sum of the constants A and C characterizes the mass adsorption on the surface in equilibrium at infinite time. B and D are rate constants describing how fast the system develops. In Fig. 7 the experimental data for the different concentrations were fitted to Equation 1. A good agreement between the measured data and the fit functions was obtained demonstrating that the initial reaction kinetic was dominated by at least two different processes. These two processes can be seen also when plotting the adsorption velocity over time (Figure 33 B). We hypothesize that the two processes involved in the initial time dependence of the frequency shift are related to monomer adsorption at the gold surface. The first process is associated to monomer adsorption at the bare resonator surface, whereas the second process corresponds to monomer deposition to S-layer patches that are already formed at the surface. A poissonian distribution describes the stochastics of both deposition processes. It might be that one of the two processes is caused by the limitation by mass transport caused by the flow cell design.

As well as the adsorption in the first few minutes, the adsorption and the change of the dissipation, which give information about the viscoelastic properties of the deposited S-layer films, were monitored over times up to two hours (Figure 33 C and D). At longer observation times, we observed deviations from the initial behaviour. While the frequency shifts tend to develop into two plateaus at around 400 and 800 MHz, characteristic oscillations in the dissipation were observed. We suggest that these results indicate the appearance of an additional structure formation processes going on in the deposited film. Since a recrystallised film is expected to be mechanically stiffer than a less ordered 'amorphous' protein film, we believe that this effect is caused by a rearrangement of monomers at the surface related to the recrystallisation of monomers into 2D protein crystal domains. We furthermore suggest that the two plateaus in the frequency shift might correspond to the formation of a monolayer and a double layer. However, to draw unambiguous conclusions about the long-term behaviour of S-layer recrystallisation, a more detailed analysis of the experimentally derived frequency shifts and dissipation data as well as their correlation will be necessary.

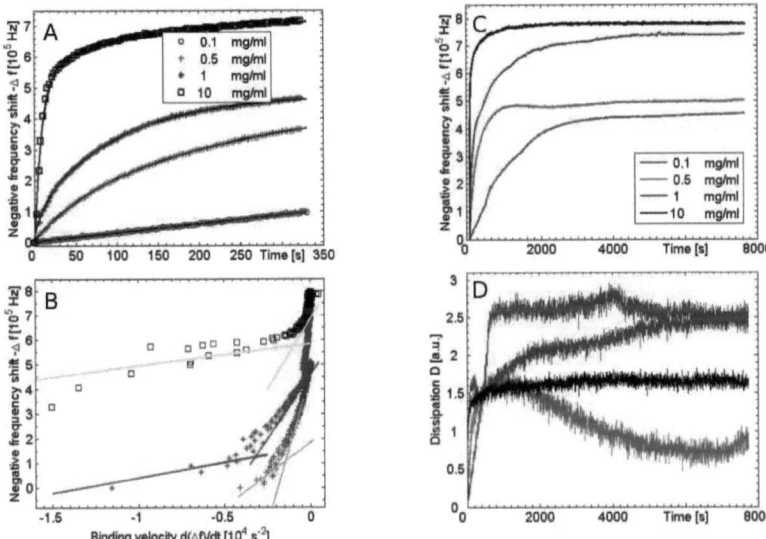

Figure 33: S layer adsorption on FBAR. (A) The adsorption of the S-layer proteins the first 5 minutes and (B) the binding velocities of the binding process during the first minutes. (C) The binding process followed over 2 hours and (D) the dissipation change of the adsorption during the first 2 hours.

5.6 Conclusion and Outlook

The FBAR used for the measurements in this chapter performed well. Their limit of detection for the surface mass in ng/cm^2 was similar to those of commercially available QCM systems and stable measurements in liquid could be performed.

The hybridisation of long, unlabelled DNA products at nanomolar sensitivity was demonstrated and the FBAR results were comparable to the SPR even in a complex sample matrix (diluted serum).

The background noise from serum was effectively minimized through use of a repelling lipoamide, Lipa-DEA. The target surface density for hybridisation with layers produced by dispensing was 25 to 50 % lower than those produced in situ. These unlabelled measurement systems seem to work well especially with longer single stranded DNA samples. The same DNA surface functionalisation will be used in Chapter 7.3 and Chapter

7.4 with a chip containing 64 pixels to demonstrate the possibility of the use of many FBARs in parallel.

The measurements of S-layer adsorption showed the possibility of the characterization of adsorption processes. By measuring the change of the energy dissipation in addition to the frequency shift, the viscoelastic properties of the adsorbates could be studied. The robustness and stability of the set-up used allowed the study of the time dependence over several hours of complex structure formation processes occuring in thin biomolecular coatings. This capability was demonstrated by the investigation of the adsorption kinetics and recrystallisation of the bacterial surface layer protein of Bacillus sphaericus NCTC 9602.

6 Lipid Bilayer and PEM[4]

Thickness shear mode (TSM) resonators in form of the QCM have been used for decades for the analysis of intermolecular interactions [167]. However, TSM resonators produced by thin-film technology, namely FBARs, capable of applications in liquid have only recently been demonstrated[36, 37, 168]. Thin film bulk acoustic resonators vibrating in longitudinal mode have been produced previously e.g. for filter applications [169], however, for use in liquid, however, acoustic resonators operating in shear mode were required, as the acoustic losses caused by longitudinal waves propagating into the liquid are too high to achieve sufficient Q-factors. Piezoelectric thin-films with the c-axis being inclined from the film normal were used to achieve sufficiently high piezoelectric shear coupling coefficients [37, 40, 41, 43-46, 170]. While the working principle of FBAR and QCM is similar, the QCM is produced in a top-down process and FBARs in a bottom-up process using thin-film technology. As a result FBARs can be made thinner, which results in a higher resonant frequency. FBARs operating from some hundreds of MHz to several GHz have been presented [39]. The small size makes it possible to integrate many resonators on a small area, which makes the FBAR a promising low-cost alternative for biomolecular interaction analysis with high throughput.

A large body of theoretical work is available for acoustic resonators. Starting with the linear dependence of a deposited mass in air [16], models were found for use of the QCM in bulk liquid [17] and for viscoelastic layers in air and liquid [18, 24, 171-173]. From these investigations it is known that the QCM responds to changes in the density, thickness, viscosity and elasticity of the adsorbate. Analogously, [174] investigated the influence of density, thickness and viscosity of adsorbates on surface acoustic wave (SAW) sensors. They used temperature-induced conformational changes of a polymer to model the SAW device behaviour. Francis *et al.* [175] specifically studied the frequency response of FBARs to protein films with thicknesses in the range of the film resonance. In these models the resonant frequency plays affects the influence of the adsorbed mass and the viscoelastic properties of the adsorbate. In order to investigate the sensor's response to adlayers with different viscoelastic properties, we used lipid vesicles as a model system. The resonant frequency response and dissipation changes upon exposure to a lipid vesicle solution has been studied intensively with QCM-D [25], and compared to other

---

[4] Parts of this chapter were published in Nirschl, M.; Schreiter, M.; Voros, J., Comparison of FBAR and QCM-D sensitivity dependence on adlayer thickness and viscosity. *Sensors and Actuators A: Physical* **2010**.

techniques such as SPR [176] and OWLS [177]. During the transition from whole vesicles to a lipid bilayer the viscoelastic properties change significantly due to the release of the solvent trapped inside the vesicles, making it interesting to compare acoustic resonators with different resonant frequencies using this model system. Furthermore, the maximum thickness of the adlayer which can be detected (i.e. the sensing depth) by the sensor is frequency dependent because of the short distance acoustic waves propagate into the sensor's surrounding (i.e. the penetration depth). In order to compare the sensing length of the QCM and the FBAR, we performed layer-by-layer deposition (LBL) of a polyelectrolyte multilayer (PEM) on QCM-D and FBAR. The adsorption of PEM films has been monitored previously with QCM [178], OWLS [179] and SPR [180]. Using an exponentially growing film, thicknesses around the expected penetration depth of several hundred nanometres were achieved [175, 181].

The purpose of this chapter is to compare the behaviour of QCM-D and FBAR during vesicle adsorption and bilayer formation and polyelectrolyte multilayer formation. Using these two techniques, acoustic resonators with a frequency range from 5 MHz up to about 2 GHz were available.

6.1 Experimental Section

6.1.1 Lipids and Vesicles

1,2-di-oleoyl-sn-glycero-3-phosphocholine (DOPC) phospholipids were purchased from Avanti Polar Lipids Inc., USA and stored dissolved in chloroform at -20°C. The chloroform was evaporated using dry nitrogen; buffer was added and the lipids extruded 31 times through polycarbonate membranes (Avestin, Canada) with a pore size of 50 nm. The buffer used was 20 mM 4-(2-hydroxyethyl)piperazine-1-ethane-sulfonic acid (HEPES, Fluka Chemie GmbH, Switzerland), 150 mM NaCl and 2 mM of $CaCl_2$. The concentration of the vesicles was 0.1 mg/ml. The vesicles were stored at 4°C in nitrogen atmosphere and used within 2 weeks.

6.1.2 Polyelectrolyte Multilayers (PEM)

Polyethyleneimine (PEI), Poly-L-Glutamic acid (PGA) and Poly-(allylamine hydrochloride) (PAH) were purchased from Sigma-Aldrich (Switzerland). Both polymers were purchased from SurfaceSolutionS GmBH (Switzerland). The polyelectrolyte multilayer built on QCM-D and FBAR was PEI-(PGA-PAH)$_n$. The polyelectrolytes were injected in time intervals of 5 minutes at a concentration of 1 mg/ml without flow and without buffer rinse between the injections as shown in [181]. The buffer used was 10 mM HEPES solution containing 100mM KCl, pH adjusted to 7.4.

6.2 Vesicle Adsorption and Bilayer Formation

Measurements on QCM-D were performed on both gold and SiO$_2$ surfaces as a reference for the FBAR measurements. On SiO$_2$, during vesicle adsorption, the resonant frequency decreased until a critical surface concentration was reached about 100 seconds after injection. At this point, the vesicles formed a lipid bilayer, which brought the dissipation closely back to zero, and the frequency increased up to about -25 Hz (Figure 34a), characteristic for a lipid bilayer. On gold, after exposure to the vesicle solution, a monotonous decrease of the resonant frequency was observed. The saturation values for both the frequency shift and the dissipation change are several times higher on the gold surface than on the SiO$_2$ surface (Figure 34 b). The adsorption on both gold and SiO$_2$ was as expected from literature [25].

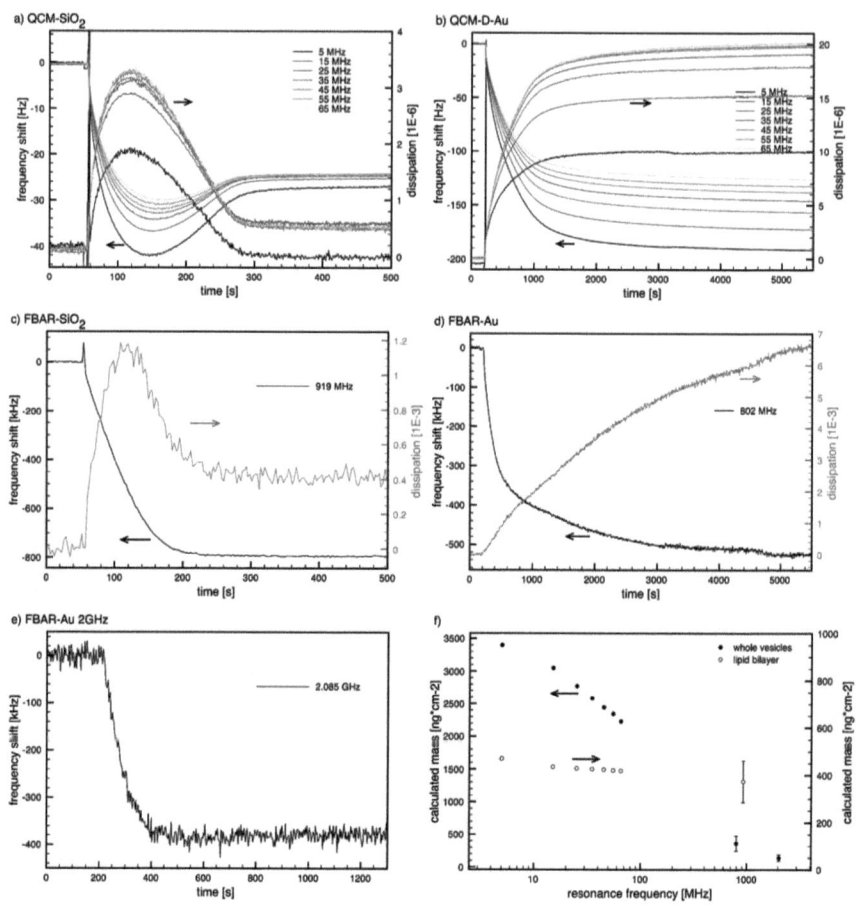

Figure 34: Comparison of vesicle adsorption and bilayer formation as measured by FBAR and QCM-D: Lipid bilayer formation on SiO2 monitored by the frequency shift and dissipation change using QCM-D (a) and FBAR (c). Adsorption of intact vesicles on gold measured by QCM-D (b) and FBAR at a resonant frequency of 800 MHz (d) and 2 GHz (e). Surface mass calculated from the frequency shift and plotted for lipid bilayer and the vesicle layer for all frequencies from 5 MHz to 2 GHz. The error bars show the standard deviation over three measurements. (f)

The measurements were conducted in the same way using the FBAR with SiO_2 and gold surfaces. On gold, vesicle adsorption analogously to the QCM-D measurements was observed. The resonant frequency decreased and dissipation increased with addition of the vesicles (Figure 34 d). On SiO_2, vesicle adsorption caused the dissipation to increase for a short time before later decreasing back to close to the original value (Figure 34 c). While this indicates a vesicle adsorption and subsequent bilayer transformation, the non-monotonic behaviour of the frequency response of the QCM-D curve cannot be seen in the FBAR measurement (Figure 34 a and c). The frequency shift at saturation was 483 kHz ± 151 kHz on gold and 592 kHz ± 139 kHz on SiO_2 surface. For the resonators with the gold surface, the vesicle adsorption was also performed with the third overtone at 2 GHz. However, due to the low Q-factor of this overtone the noise of the dissipation was higher than the signal expected from vesicle adsorption, therefore the adsorption could only be observed from the change in resonant frequency. The frequency shift at saturation was 305 kHz ± 105 kHz, lower than the shift of the fundamental frequency, although from the simulation the overtone would be expected to have a higher mass sensitivity (Figure 34 e, Figure 37 a).

The frequency shifts at saturation for all measurements were converted into surface mass using the Sauerbrey equation for the measured values from QCM-D [16] and the Mason model for the values from the FBAR (Figure 34 f). It can be seen that the calculated mass decreases with higher frequencies in case of the vesicles while it nearly stays constant for the lipid bilayer. Interestingly, while the Sauerbrey mass measured using QCM-D is higher for the vesicles than the lipid bilayer as expected, it is the other way round on the FBAR.

6.3 Polyelectrolyte Multilayer (PEM)

The deposition of the polyelectrolyte multilayer on QCM-D was followed by a monotonic decrease of the resonant frequency with increasing layer number. On the other hand, the dissipation increased after PGA and decreased after PAH addition in agreement with the literature [181-183]. This indicates the addition of PAH as the uppermost layer yields the formation of a more rigid film (Figure 35a). The PEM film was built in the same manner on the FBAR as on the QCM-D. It was observed that the frequency decreases after addition of PGA until the 7th layer pair. From the 8th layer pair on, the frequency increases again in comparison to the previous PGA injection. On the contrary, after the addition of PAH, the frequency increases rapidly to a much smaller frequency shift. The dissipation was high after adding PGA and lower after adding PAH (Figure 3b), which is the opposite of the QCM-D measurement.

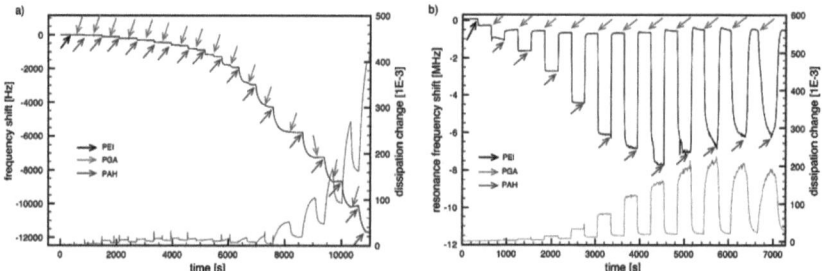

Figure 35: Polyelectrolyte multilayer formation on QCM-D (a) and FBAR (b) Black arrows indicate the injection of PEI, red arrows injection of PGA and blue of PAH. While the frequency decreases monotonously on QCM-D, on FBAR large jumps appear at every injection.

The large signals obtained by changing from PGA to PAH and vice versa gave rise to the question of whether the ionic state inside the PEM can also be changed by another type of ion. This would open up the possibility of using this device as an ion sensor.

To answer this question, ferrocyanide was rinsed over a PEI-(PGA-PAH)$_{12}$-PGA PEM. As with adding PAH, large positive frequency shifts were obtained. The frequency shift was greater than 10 MHz with the ferrocyanide in the micromolar range. Notably the points follow a straight line if plotted on a log-log plot (Figure 36). This indicates that the adsorption process cannot be represented by only a simple molecular binding process (e.g. a Langmuir adsorption) but another effect, for example the diffusion of the ions into the PEM might play a role.

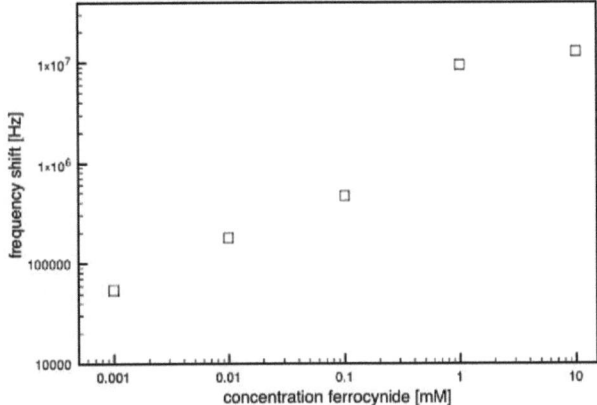

Figure 36: Titration curve of ferrocyanide on a PEI-(PGA-PAH)$_{12}$-PGA polyelectrolyte multilayer. The points for the frequency shift are close to being on a line in this log-log plot.

6.4 Discussion

The measurements showed that vesicle adsorption and lipid bilayer formation, and the LBL deposition of PEMs could be observed on FBAR, and that significant differences were found for different resonant frequencies.

We assume that the QCM and the FBAR have identical surfaces (gold or SiO_2) as they were produced in similar processes, therefore the influence of effects like surface roughness or other surface properties that might lead to differences in the adsorption process or directly cause a frequency shift are neglected.

Instead, we suggest, that only the increased resonant frequency of the FBAR causes the deviation from the QCM-D, which operates at lower frequencies. At a higher resonant frequency, the penetration depth δ is lower according to the estimation $\delta = \sqrt{\frac{\eta}{\pi \rho f_0}}$, where η is the viscosity and ρ the density of the surrounding medium. δ determines the length scale over which resonator is sensitive.

Normally, for thin adlayers, the penetration depth is determined largely by the solvent. However, for thick films and high frequencies the adlayer might be thicker than δ [175]. The viscosity and density of the layers can be estimated from the QCM-D measurements assuming that the influence of the elasticity can be neglected ($\eta_{vesicles} = 1.14*10^{-3} \frac{kg}{ms}$ and $\rho_{vesicles} = 1030 \frac{kg}{m^3}$).

To obtain a rough estimate of the penetration depth it is assumed that the viscosity is frequency independent which is unlikely to be the case. The resulting penetration depth is then would be 265 nm at 5 MHz, 69 nm at 75 MHz and at the FBAR operating frequency of 800 MHz it is around 21 nm. For the overtone at 2 GHz it would be only 13 nm. In the QCM-D measurements the penetration depth is higher than the thickness of both the lipid bilayer and the whole vesicles. Therefore the sensor response is related to the whole vesicle including the coupled water. As such, the calculated mass is much higher than for the case of the lipid bilayer [176]. On the FBAR, however, the penetration depth is smaller than the 50 nm diameter of the vesicles. As a result, the vesicles are only partially penetrated by the acoustic waves and as such only part of the vesicles contribute to the frequency shift.

The size of the frequency shift is determined by both the fraction of the vesicle and coupled water penetrated by the acoustic wave and by the packing density of the adsorbed vesicles.

In contrast, the lipid bilayer instead is completely contained within the sensing depth; therefore all of its mass contributes to the frequency shift. This explains why we obtained a larger mass for the bilayer than for the vesicles at 800 MHz (Figure 34f). The dissipation change still shows the vesicle rupture event, probably due to the change in viscoelastic properties of the part of the vesicles penetrated by the acoustic waves (Figure 34c and d and Figure 37c).

The influence of the penetration depth helps also to explain the frequency shift obtained at 2 GHz. The simulated mass sensitivity is higher for the mode at 2 GHz than for the one at 800 MHz, but the frequency shift of the adsorbed vesicles was smaller for the overtone at 2 GHz. This is because the penetration depth is 13 nm at 2 GHz, which leaves an even larger part of the vesicles "unseen" by the resonator, and only a small part contributes to the resonant frequency shift. In order to better understand the consequences of the altered penetration depth at different frequencies we have simulated the expected frequency shifts upon adsorption of homogeneous layers with different thicknesses. Figure 4a shows the simulated frequency shifts of a viscoelastic layer with viscosity and density of $\eta_{vesicles}$ and $\delta_{vesicles}$. The simulation indicates remarkable differences between QCM-D and FBAR, e.g. the critical thickness at which the frequency shift has a maximum is very different. Another situation when the sensor response depends on the resonant frequency is related to the influence of the viscoelastic properties of the adsorbate. The frequency is not only influenced by the mass attachment itself but also by the viscosity and elasticity of the adsorbed layer. The frequency shift and dissipation caused by a viscoelastic layer deposited on an acoustic resonator operated in bulk liquid has been described as the combination of three contributions: The bulk liquid [17], the mass of the adsorbate [16] and the viscoelastic properties of the adsorbate [172]. The viscoelastic properties contain the viscosity, which cannot be assumed to be constant over the broad frequency range from MHz to GHz. A generalised model for the complex shear modulus has been introduced in the form $G^* = \dfrac{G}{(1-\dfrac{i}{\omega\tau})^b}$ by [172]. The complex shear modulus was modelled as the combination of a distribution of relaxation times τ characterised by the factor b and the resonant frequency ω. Following this model, we suggest that for the adsorbed vesicles the distribution of relaxation times allows excitation at low frequencies while the relaxation times are too high to allow sufficient excitation at the higher resonant frequency of the FBAR. In this way, the contribution of the viscoelastic properties to the frequency shift is small compared to the contribution of the mass attachment. The real and the imaginary part of G^* have a different frequency dependency depending on the distribution of relaxation times, so that the influence on the dissipation can be high while it is low on the resonant frequency. Thus, the transition from whole vesicles to lipid bilayer cannot be seen from the frequency shift, but only from the difference in dissipation between the whole vesicle, which is the at least partially "seen" by the resonator, and the lipid bilayer. The fact that frequency shift and dissipation do not necessarily behave in an analogue manner can also be seen from the vesicle adsorption on QCM-D, where the frequency shift at saturation decreases while the dissipation increases at higher frequencies (Figure 2b).

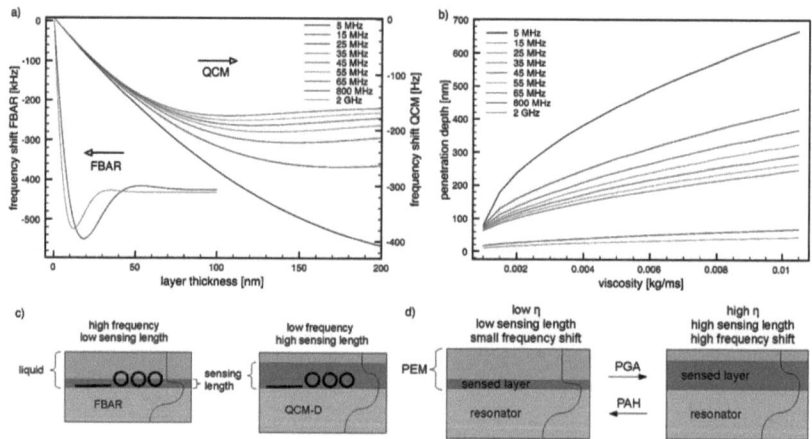

Figure 37: a) Calculated adsorption of a viscoelastic layer on QCM and FBAR for different frequencies. The penetration depth is significantly different between the two techniques and for different resonance frequencies. The sketch in c) illustrates the impact of the different penetration depths on the vesicle adsorption: The whole vesicles are penetrated by the acoustic waves on QCM-D while the sensing depth is shorter than the vesicle diameter on FBAR. b) The sensing depth (i.e. layer thickness at maximum frequency shift) for different viscosities of a 150 nm thick adlayer. With lower viscosity and higher frequencies the sensing depth decreases. d) The sketch illustrates the mechanism causing the high frequency jumps on FBAR, the change in the viscosity of the PEM moves mass in an out of the sensing depth.

Further differences between QCM-D and FBAR could be seen from the polyelectrolyte multilayer deposition. QCM-D, increasing film thickness caused a decrease in frequency shift. On FBAR, the frequency decreased only upon addition of PGA. After adding PAH the frequency rapidly recovered. The dissipation change shows alternately a highly and a weakly dissipative layer in contrast to the QCM-D measurement (Figure 35). From previous work [184] it is known that the PEMs viscosity depends on which polymer was added last. This change in viscosity has a large influence on the penetration depth. In Figure 37 b) the sensing depth is plotted for different viscosities for a 150 nm thick adlayer, where the thickness at which the maximum frequency shift occurred was taken to be the sensing depth. This indicates that the sensing depth of the FBAR is smaller than the thickness of the adlayer. Accordingly, if changes in the viscosity alter the sensing depth the proportion of the PEM, which is sensed by the resonator, will also be different. The part of the adsorbate moved out of the sensing depth does not contribute to the resonant frequency and dissipation change. By changing the viscosity and thus the sensing length within the PEM, the mass located between the two sensing lengths is alternately sensed or not sensed by the resonator, causing the big changes in frequency and dissipation. The sketches in Figure 37 d) illustrate this process.

From Figure 35 b) it can be seen that the maximum frequency shift, which we assume to be the penetration depth, is reached after the 5th layer for the PGA and for the 7th layer for the PAH. From the QCM-D measurements the corresponding thickness was calculated to be 80 nm for the PGA and 210 nm for the PAH. Assuming that the PEM is a homogeneous stack with a density of $1100 \frac{kg}{m^3}$ for both polymers we receive an estimated viscosity for the PEM of $\eta_{PGA} = 17.7*10^{-3} \frac{kg}{ms}$ if PGA is the last layer and $\eta_{PAH} = 122*10^{-3} \frac{kg}{ms}$ if PAH is the last layer.

6.5 Conclusions

In this chapter, we have compared the sensor response of the FBAR with QCM-D upon adsorption of vesicles and the formation of a lipid bilayer, and the LBL deposition of PEMs. We have shown that there are significant differences between the two different types of acoustic resonators and between different resonance frequencies. We have suggested that in the measurements we have done, the decreased penetration depth and the smaller influence of viscoelastic properties at higher frequency caused the unexpected results. Simulations using the Mason model confirmed this hypothesis.

As a result, the properties of the adsorbate (e.g. thickness, viscoelastic properties) have to be taken into consideration when selecting the resonant frequency of acoustic resonators as mass sensors in order to be able to efficiently sense the adsorbate. While a higher resonant frequency makes the device thinner and more sensitive, the frequency has an upper limit where the sensing length of the device becomes too short, e.g. because the assay to be used has a thickness in the range of the sensing length. However, in an application where a high sensitivity to changes in the viscosity is desired the operating frequency might be selected in a range where the adsorbate has a thickness near the sensing length, as was the case for the measurements presented here.

7 CMOS-integrated FBAR Array for Specific and Selective Multiplexed Detection of DNA in Buffer and Diluted Serum[5]

In previous chapters the FBARs were read-out using a network analyser. This is costly and makes it difficult to measure many pixels at simultaneously. In order to obtain responses from a large number of pixels, a small and cheap read-out circuit specifically designed for the purpose of determining the resonant frequency of multiple resonators is necessary. Interface circuits for the QCM for application in air and liquid have been developed and optimized for decades [185]. The development of a read-out circuit for FBARs, however, has different prerequisites than for the QCM because of the generally lower quality factors of the FBAR, and the larger variation of the resonant frequencies of the resonators originating from the FBAR production process. While oscillator based read-outs were previously presented [186, 187], an impedance-based read-out is more robust especially in a liquid environment. Schneider et al. [188] presented a design utilizing a direct digital synthesis, where a test signal is generated using a digital signal processor, D/A-converted and the corresponding power is measured. As an additional advantage of this method, not only the resonant frequency, but also the changes in the energy dissipation can be determined. This can be useful to determine viscoelastic properties of adsorbates [189], [178].

This direct digital synthesis leads to a highly complex readout circuitry and the maximum frequency band is quite limited. The importance of a simple readout is emphasized when the impedance analysis is utilized for an integrated sensor matrix. A novel, simple, but accurate impedance change detection method was developed for FBAR resonators. The core part of the analyser is a ring oscillator based voltage-controlled oscillator (VCO). The VCO frequency is swept over a large frequency band and the FBAR

---

[5] Parts of this chapter were published in Nirschl, M.; Rantala, A.; Tukkiniemi, K.; Auer, S.; Hellgren, A.-C.; Pitzer, D.; Schreiter, M.; Vikholm-Lundin, I. CMOS-Integrated Film Bulk Acoustic Resonators for Label-Free Biosensing. *Sensors* **2010**, *10*, 4180-4193 or Tukkiniemi, K.; Rantala, A.; Nirschl, M.; Pitzer, D.; Huber, T.; Schreiter, M. Fully integrated FBAR sensor matrix for mass detection. *Procedia Chemistry* **2009**, *1*, 1051-1054
CMOS read-out was designed and provided by the group of Kari Tukkiniemi (VTT Finnland)

impedance response is acquired from the VCO output. A dedicated algorithm resolves the resonant frequency of the FBAR. [190]

In this chapter, we present arrays with 64 acoustic resonators on one chip, which are monolithically integrated into a novel impedance-based CMOS. The sensitivity of the device is determined and the usability for label-free DNA detection is demonstrated.

7.1 Experimental Section

7.1.1 CMOS-integrated FBAR

The FBARs were processed on CMOS wafers similar to the ones described in Chapter 3.2. Instead of being produced on standard silicon wafers, they were back-end processed in arrays of 4x16 pixels on active 0.35 μm CMOS wafers. A dedicated read-out circuit was located under each FBAR pixel. This circuit included an interface block which provided communication with the system level, a voltage-controlled oscillator (VCO), a local control which interprets the system level commands and controls the VCO, and a frequency meter (digital counter). The resonant frequency acquisition is performed as follows: A value for the VCO control is calculated at system level and applied to the corresponding pixel. The corresponding frequency is obtained by integrating the VCO output with the digital counter.

At the beginning of the analysis the operating point for each FBAR pixel was set. For this, the control for the VCOs was swept along a range around the resonant frequency while the output frequency in recorded. The control voltage versus frequency curve acquired carries information about the electrical impedance seen by VCO (i.e. the FBAR sensor) [190]. An empirical algorithm was developed to detect the operation point corresponding to the parallel resonant frequency of the resonator.

Once the operation point for all pixels has been acquired, the control value for VCO was fixed during the measurement. During the measurement the output frequencies of the VCOs were continuously measured and the observed frequency deviations reflect the changes of the FBAR resonant frequency.

The finished chips were then glued onto a printed circuit board, wire bonded and sealed. About a dozen of chips with all 64 pixels working were available from one 6" wafer. A SEM picture of one of the resonators can be seen in Figure 38 a). Figure 38 b) shows the complete chip with 64 pixels under a quartz crystal. The cartridge (Figure 38 c) contained a flow cell (about 10 μl volume) with a simple inlet and outlet, which allowed manual injections of the liquids using a syringe. A minimum amount of 500 µl was injected to ensure complete exchange of the liquids.

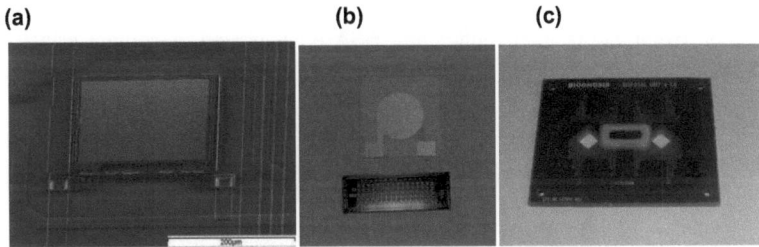

Figure 38: (a) A scanning electron microscope picture of one of the FBAR pixels. The pixel size is 200 µm x 200 µm. (b) FBAR chip under a QCM crystal. While the physical working principle is similar, the FBAR allows integrating 64 pixel on about the same area like a QCM crystal. (c) The CMOS-integrated FBAR packaged on a credit-sized board, which allows a very simple handling of the sensor.

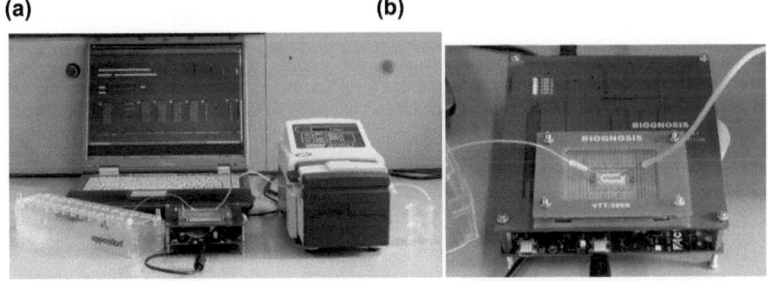

Figure 39: (a) The set-up for measurements with the CMOS-integrated FBAR: Fluid is pumped using a peristaltic pump (on the right) over the read-out (middle) into a waste vessel. (b) the read-out in detail: The FBAR cartridge is mounted on two PCB boards; the lower one contains the FPGA board for the read-out electronics.

7.1.2 BSA Measurements

A 10 MHz QCM Biosens system (Biosensor Applications AB, Sweden) was read-out at its fundamental frequency. BSA at concentrations ranging from 1 - 1000 µg/ml was deposited on the gold surfaces of the devices while recording the change in refractive index and resonance frequencies for the SPR, QCM and FBAR, respectively. The lowest BSA concentration was added after a stable baseline had been recorded for 1 minute for SPR and at least 5 minutes for QCM and the FBAR. Buffer and BSA in increasing concentrations were injected alternately for 5 minutes each. The flow speed of the SPR was 20 µl/min, and the liquids were injected onto the QCM and FBAR using a syringe and without flow. In all experiments, both buffer and BSA each stayed in contact with the sensor for 5 minutes. The measurement was performed at 25 °C on SPR and the passive FBAR. Room temperature was used for the CMOS-integrated FBAR and for the QCM.

7.2 Mass Sensitivity Comparison obtained with FBAR, SPR and QCM

In order to experimentally determine the mass sensitivity of the FBAR, reference measurements were conducted comparing the FBAR, QCM and SPR. All sensors were exposed to BSA in concentrations ranging from 1 - 1000 µg/ml. All three sensors show the same trend, i.e. an increase in response with increased concentrations. Figure 40 shows the resulting titration curve of the CMOS-integrated FBAR, QCM and SPR.

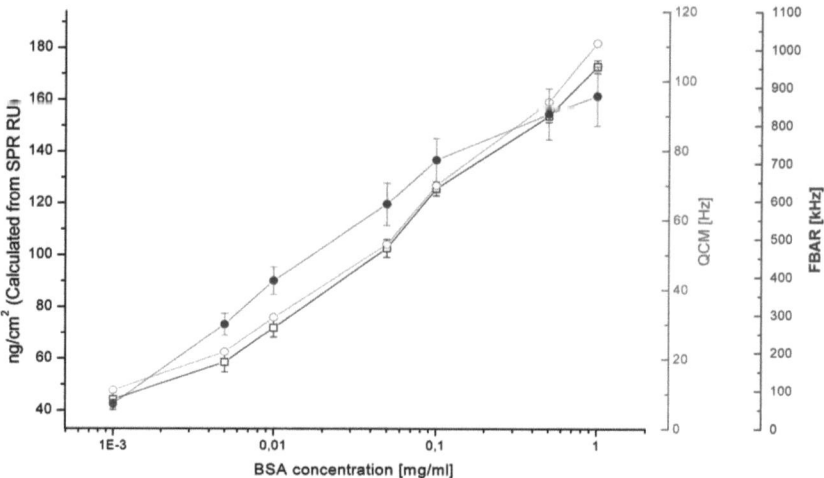

Figure 40: Titration curves for bovine serum albumin as measured with QCM (¡), CMOS-integrated FBAR (l) and SPR (o). The curve for the FBAR measurement is the average of 10 pixels, the SPR represents 4 channels; the error bars show the standard deviation.

The SPR results were used to determine the adsorbed surface mass obtained for each BSA concentration because the response in resonance units can be easily converted into surface mass: One resonance unit corresponds to a surface mass of 0.1 ng cm^{-2} [58]. With the mass of BSA adsorbed to the surface known for each concentration, the corresponding frequency shifts of the FBAR could be calibrated, ie. the mass sensitivity in terms of frequency shift per adsorbed mass calculated. The same calculation was done for the QCM in order validate the results as the mass sensitivity for the QCM, in contrast to the FBAR, can be calculated using the Sauerbrey equation. Table 7 summarized the mass sensitivity, the noise of the measurements and the resulting limit of detection. The experimentally obtained mass sensitivity of the QCM is about three times higher than the value obtained using the Sauerbrey equation. This result is as expected because the QCM unlike the SPR does not only sense the adsorbed protein but also the liquid coupled to it [191]. This suggests that this experimental method is applicable for determination of the mass sensitivity of the FBAR. The mass sensitivity of the FBAR is nearly two orders of magnitude higher than the sensitivity of the QCM. This is in agreement with what has been shown previously [192]. However, in order to properly evaluate the FBAR sensor performance, the noise level, also higher by nearly 2 orders of magnitude, must be taken into account [136]. The resulting limit of detection (LOD), the smallest surface mass that can be detected, is defined as the frequency shift per surface mass divided by three times the frequency noise over 10 measurements. The LODs were found to be similar for QCM, the passive and CMOS-integrated FBARs. The SPR has a 7-25 fold lower LOD than the acoustic resonators investigated.

The standard deviation of measurements between several FBARs is greater than the noise of the resonators and also larger than the standard deviation of the measurements obtained with SPR. The curves recorded with QCM more closely follow the one recorded by SPR, while the curve from FBAR shows a slightly different behaviour. However, due to the presence of visible air bubbles in the flow cell used with the FBAR, and frequency jumps shown by some resonators during the injection of liquids, the main influence for the differences in the signal and the higher standard deviation might be air bubbles in the vicinity of the resonator surface. The results are therefore quite promising considering that Biacore 3000 uses a highly developed fluidic system whereas no emphasis was put on the FBAR fluidics and the FBAR could be further improved.

	SPR	QCM	Passive FBAR	CMOS-integrated FBAR (best/average)
Mass sensitivity	10 RU cm²/ng [58]	61.1 ppb cm²/ng (measured) 22.7 ppb cm²/ng (Sauerbrey [16])	5.63 ppm cm²/ng	7.13 ppm cm²/ng
Noise level (3s)	0.63 RU	23 ppb	2.3 ppm	3.0 / 10.8 ppm
Mass resolution (LOD)	0.06 ng/cm²	0.38 ng/cm²	0.41 ng/cm²	0.42 / 1.5 ng/cm²

Table 7: Mass sensitivity, frequency noise and limit of detection for SPR, QCM, the passive and the CMOS-integrated FBAR. The mass sensitivity for the acoustic resonators is relative to their resonant frequency. For the CMOS- integrated FBAR, the noise and LOD from the best pixel and the average over a 64-pixel array are shown.

7.3 Multiplexed DNA Measurement

In order to demonstrate that a large number of pixels can be used simultaneously, a pattern was spotted onto the sensor array. Figure 41 a) shows the 4x16 pixel array after the S-S-CALCA/Lipa-DEA solution was spotted on some pixels, which appear in black. The pixels that appear white are the blank gold surfaces and were coated in a subsequent step with S-S-PTGS2/Lipa-DEA that was used as a reference coating. The arrangement of the pixels represents the letters DNA.

Figure 41 b) shows the sensor response of all 64 pixels at saturation after hybridisation with the complementary DNA (CALCA). A frequency shift of -270 ± 80 kHz was observed for the pixels functionalized with the probes after a 5 minutes interaction period. No frequency shift (2 ±17 kHz) could be observed for the pixels passivated with Lipa-DEA.

Figure 41 c) shows the time resolved frequency shifts of 50 pixels. The remaining 14 pixels on the chips are not shown because their resonant frequency showed high positive frequency shifts upon sample injection. Just as in the measurement with BSA, air bubbles could be seen in the flow cell, which are likely to be the reason for the failing resonators.

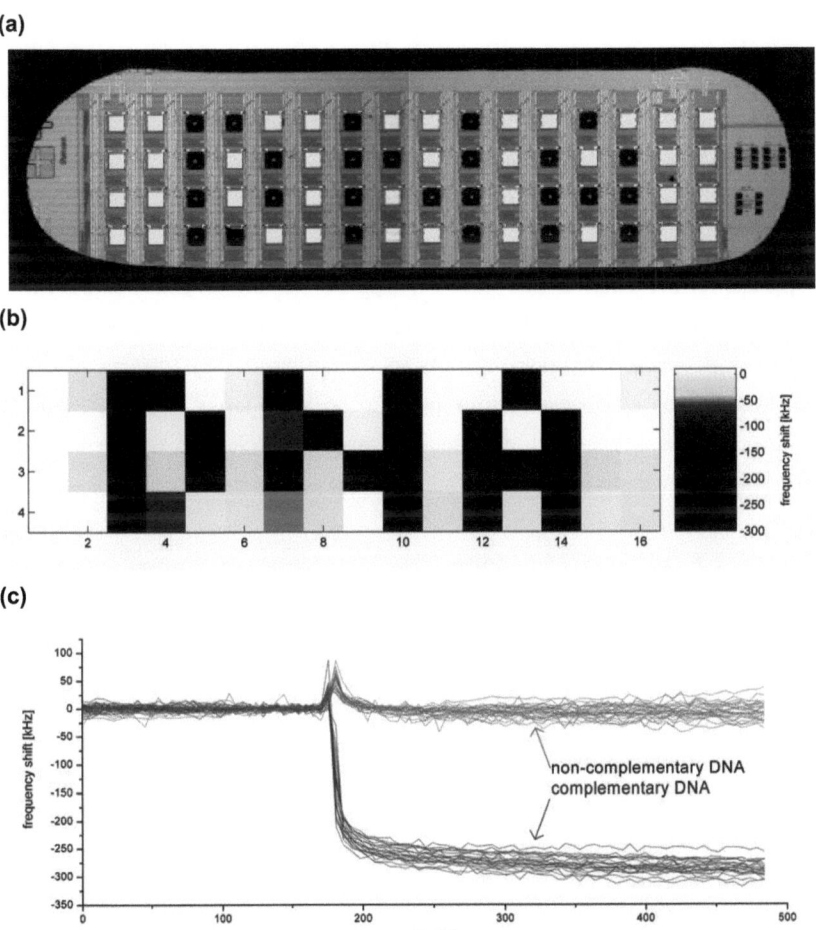

Figure 41: (a) The FBAR array with some of the pixels functionalized. The black squares are drops of liquid containing probes and Lipa-DEA. The squares appearing in white are uncovered gold pixels. (b) Frequency shift of all 64 pixels after the addition of the complimentary DNA at saturation. (c) The frequency curves of 50 selected pixels. The complementary DNA was added at t=180s.

7.4 Multiplexed Measurement of PCR Amplified Products in Buffer and Serum

Resonator arrays with three different functionalisations were prepared: S-S-PTGS2/Lipa-DEA, S-S-CALCA/Lipa-DEA and Lipa-DEA only. Figure 42 shows the different frequency responses obtained for pixels functionalized with S-S-CALCA, S-S-PTGS2 and blocked with Lipa-DEA upon interaction with complementary DNA. When the complimentary CALCA PCR product at a concentration of 1 µM was added (t=5 min) a change in resonant frequency corresponding to -332 ± 3 kHz was observed. The response was obtained only for the resonators functionalized with S-S-CALCA/Lipa-DEA, while the frequency shifts for the pixels coated with only Lipa-DEA or S-S-PTGS2/Lipa-DEA were below noise. A frequency shift of -130 ± 10 kHz was observed for the S-S-PTGS2/Lipa-DEA functionalized pixels after adding a 1 µM solution of the PTGS2 PCR product (t=30 min). The hybridization efficiency was previously found to be lower for PTGS2 than for CALCA [122]. Again, as expected, no frequency shift was observed for the Lipa-DEA passivated pixels or for the pixels functionalized with the non-complimentary probes.

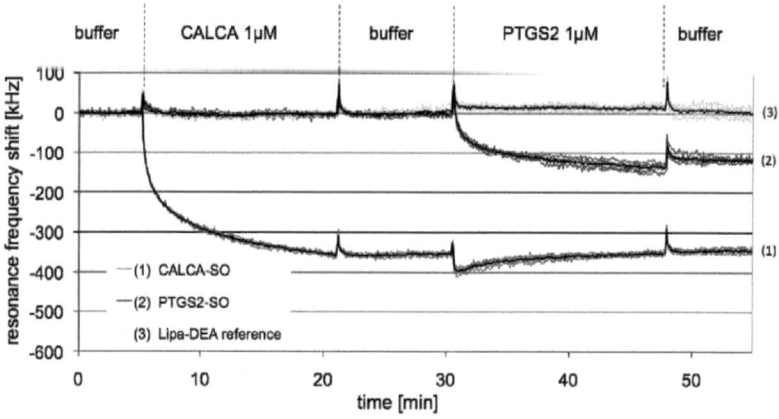

Figure 42: The frequency response of selected pixels of one FBAR chip: The red curves correspond to surfaces functionalized with S-S-CALCA/Lipa-DEA, the blue curves to surfaces functionalized with S-S-PTGS2/Lipa-DEA and the grey curves to surfaces passivated with Lipa-DEA only. The number of replicas is 5 for each type of functionalisation; the black curves are the average of the 5 curves.

For the measurement in serum, a chip was functionalized with S-S-CALCA/Lipa-DEA and S-S-PTGS2/Lipa-DEA. Figure 43 shows the FBAR response upon addition of diluted serum spiked with the complementary PCR products. A large frequency shift of -360 ± 20 kHz was observed when pixels functionalized with the complementary sequence were allowed to interact with a serum solution spiked with a CALCA PCR product to a concentration of 1 µM (1:100). The frequency shifts of the pixels functionalized with the non-complementary probes were below noise. The surface was then regenerated. This data was removed from the measurement curve for clarity because the viscosity changes caused large frequency jumps. The resonant frequency returned to the value of the buffer baseline showing that the hybridized DNA strand had been effectively removed. Upon addition of the PTGS2 PCR product at a concentration of 1 µM, a frequency shift of -130 ± 20 kHz was obtained, smaller than that of the CALCA. This agrees with the measurements in buffer. Again, no hybridization occurred on the pixels functionalized with S-S-CALCA/Lipa-DEA showing that there was no binding of non-complementary strands.

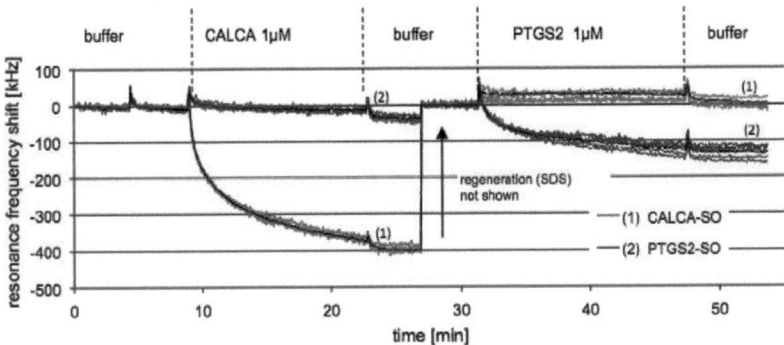

Figure 43: The frequency response of selected pixels of one FBAR chip: The red curves (5 pixels) show responses for surfaces functionalized with S-S-CALCA/Lipa-DEA and the blue curves (5 pixels) for surfaces functionalized with S-S-PTGS2/Lipa-DEA. The black curves represent the average of the 5 curves.

Additionally, a titration curve was recorded for the S-S-CALCA/Lipa-DEA pixels in buffer and serum (1:100) for the complimentary PCR product from 10 pM - 1µM. The surface was regenerated by washing the layer with a SDS-NaOH solution between each different concentration. The frequency shifts at saturation are shown in Figure 44 for buffer and serum for four selected pixels. The error bars show the standard deviation of the frequency shift of four pixels. The titration curves of the measurements in buffer and serum are similar except at the highest concentration (1 µM). Layers of hybridized DNA are known to form solvent rich layers [193]. Therefore differences in the viscoelastic properties of a layer containing high amounts of serum or buffer may be the cause of the difference in the signal at 1 µM. However, it requires future investigation to find out if this effect is a problem for quantitative measurements.

There is no visible non-specific adsorption at the lower concentrations (i.e. 1 nM and lower). This suggests that the functionalisation successfully suppresses the binding of serum.

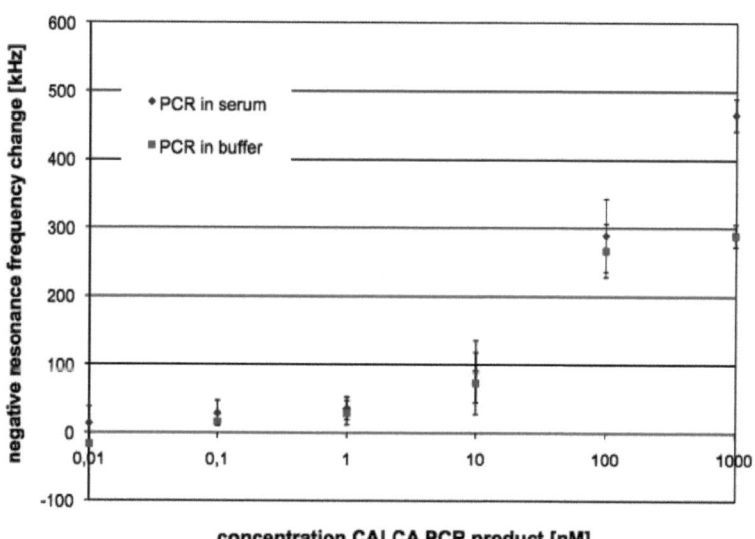

Figure 44: Titration curve for hybridisation of a CALCA PCR product in buffer (n) and serum (u) (1:100) with a surface layer of S-S-CALCA/Lipa-DEA. The error bars show the standard deviation over 4 pixels. All points were recorded on the same chip.

7.5 Combination of Compartments and Piezo Dispenser for Label-free Biosensing

The small size of the FBAR and the FBAR read-out, together with the low volume of the liquid samples makes is tempting to perform measurements on the FBAR during the spotting. With such a system, label-free measurements could be made in nanolitre volumes with high throughput and a minimum of sample consumption. Complete mixing and titrations could be performed on chip.

Figure 45 right shows the FBAR read-out inside the piezo dispenser. However, it was found that a resonator that is covered with a spot as shown in Figure 41 a) and Figure 46 A) is so strongly damped that the resonance disappears in the impedance spectrum. While an FBAR resonator works fine in air, and is only slightly damped in bulk liquid, a small droplet prevents it from resonating. One explanation for this phenomenon might be the surface tension of the liquid droplet that prevents the resonator from operating. A second limitation of this way of spotting is that it is difficult to spot several times on the same pixel to mix liquids or perform titrations because the maximum liquid volume is limited. A lower limit is about 1 nl, the minimum volume that ensures the whole sensor area is fully covered. When the droplet is too big it can move sideways away from the resonator and intermix with the neighbouring pixels. To avoid this problem compartments consisting of a photo resist (SU-8) can be built around the resonator (Figure 46 B). Figure 45 shows one of the resonators surrounded by a compartment ring. With this configuration it is possible to spot on previously functionalised pixels, and hence, spot several times on the same resonator (Figure 46 C).

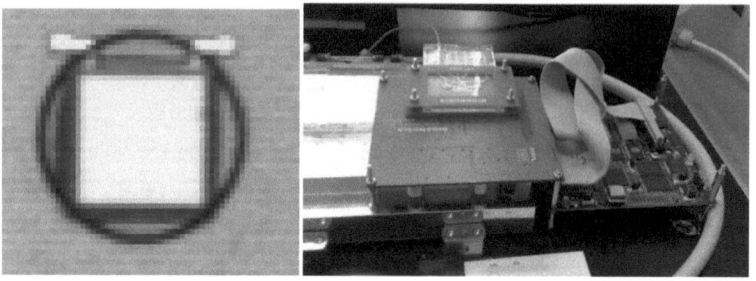

Figure 45: Left: Picture of the compartment (round) around the FBAR pixel (squared) Right: The modified read-out inside the picolitre dispenser.

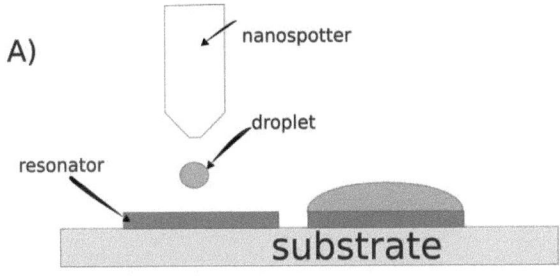

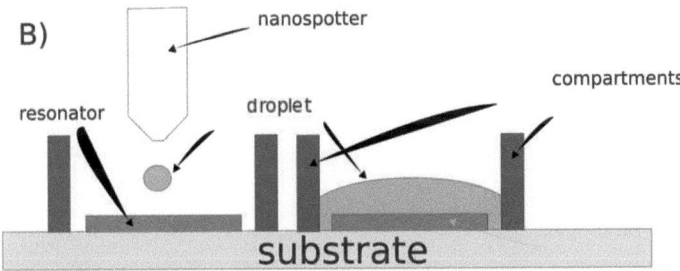

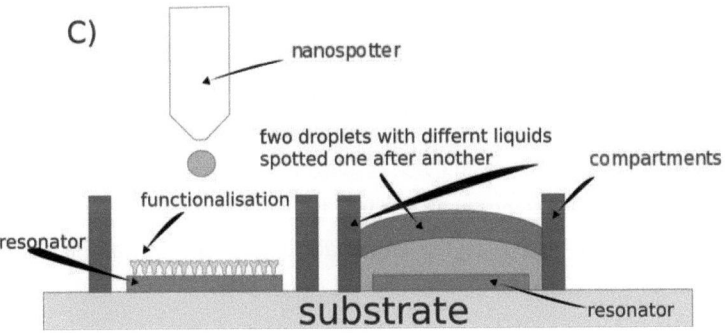

Figure 46: Sketch of the FBAR integrated into a nanospotter: Because without compartments it is difficult to spot on the resonators (A) compartments are added to make the spotting easier (B) and to enable to spot multiple times on the same resonator (C)

In a first measurement, buffer solution was spotted in different volumes (Figure 47). The number of drops per pixel was varied from 15 drops to 120 drops. As one drop corresponds to about 300 pl, this shows that the volume could be varied from about 5 nl to 36 nl.

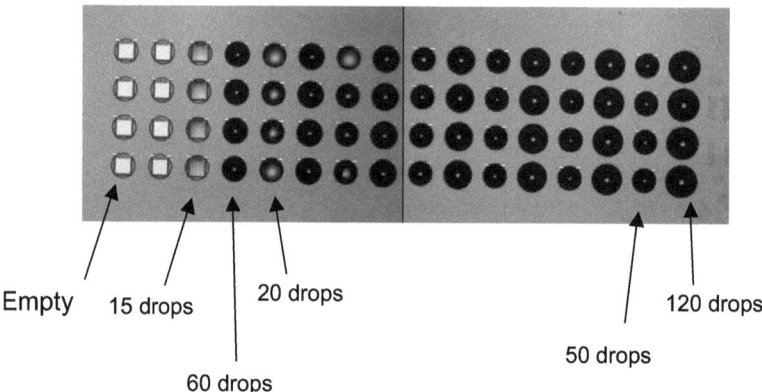

Figure 47: Buffer solution spotted on a chip with FBAR resonators and compartments with varied drop size.

Unfortunately, there were no working chips with compartments available, so spotting tests had to be performed on dummy chips that had the resonator structures and the compartments but were not electronically functional. However line arrays with compartments were available (Figure 48 a) which could be readout using the network analyser as described in Chapter 3.2. With these resonators the electrical impedance was measured as a function of the frequency in air and with the liquid droplet on top (Figure 48 b). The resonant peak with the droplet on top was smaller than in air but was still clearly visible and is likely to remain usable for biochemical measurements.

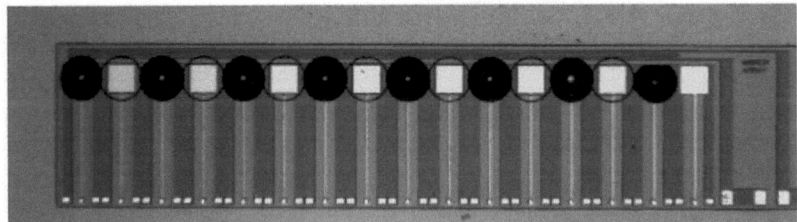

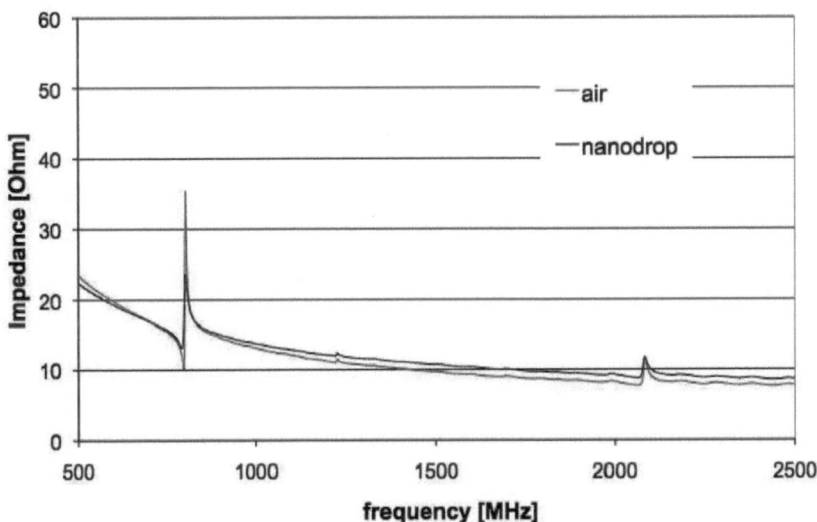

Figure 48: (a) Line array of FBARs with compartments. (b) Electrical characterisation of a resonator in air (red line) and the same resonator with a nanodrop on top (black line).

Although more time and resources were not available for experiments over and above these preliminary measurements it was shown that the FBAR could be used to conduct measurements in real-time in nanolitre volumes.

7.6 Conclusion

In this chapter, arrays of acoustic thin-film resonators integrated in an innovative CMOS read-out circuit were presented. Due to the simple and efficient read-out it was possible to integrate multiple acoustic resonators on one chip using standard CMOS technology and thin-film processing. The chip presented comprised 64 resonators on an area smaller than 1 cm^2. The sensor array showed a limit of mass detection equal to the non-integrated FBAR and the QCM. In the current state of development, SPR had a 10-fold lower detection limit, however, the FBAR has several advantages over the optical technologies: The read-out set-up is cheaper as it does not require any optical equipment. The smaller size results in lower sample consumption and higher throughput by allowing highly multiplexed measurements.

In a measurement of DNA detection in buffer, the sensor worked robustly in a liquid environment. The detection of DNA in diluted serum also successfully demonstrated that measurements in crude samples could be performed with the FBAR. Functionalised FBARs were shown to be capable of differentiating two different DNA sequences in human blood serum (1:100) without significant non-specific binding. The results achieved with the CMOS-integrated FBARs were comparable to those obtained with the passive FBAR in Chapter 5.4.

Preliminary experiments showed that it should be feasible to measure adsorptions on the FBAR during the spotting process, making it possible to reduce the required liquid volume significantly down to the range of one nanolitre per pixel. However, further experiments have to be conducted to fully evaluate the performance of this liquid system.

8 Conformational change of Calmodulin[6]

8.1 Introduction

Previous measurements using the FBAR as a biosensor for protein or DNA detection (e.g. Chapter 5 in this thesis) were based on the principle that the FBAR is sensitive to changes in adsorbed mass on the sensor surface. In Chapter 6, the highly sensitive reaction of the FBAR to changes in the viscoelastic properties of the adsorbate (i.e. a polyelectrolyte multilayer) was described.. In this chapter we investigate the FBAR response to changes in the conformation of a protein. As an example system we measure the binding of calcium and the CaMKII peptide to calmodulin. Because the mass of the calcium is too small to be detected, the conformational change caused by the binding process is measured by monitoring the resonant frequency and the motional resistance of the FBAR. The resonant frequency is, under certain conditions (e.g. where the adsorbed layer is thin compared to the resonator), a measure of the amount of mass coupled to the sensor while the motional resistance is largly influenced by the viscoelastic properties of the adsorbate.

The frequency shift measured during the calcium adsorptions was found to be strongly dependent on the surface concentration of the immobilized calmodulin, indicating that the measured signal is significantly influenced by the amount of water inside the calmodulin layer.

By plotting the measured motional resistance against the frequency shift, a mass adsorption can be distinguished from processes involving measurable conformational changes. With this method three serial processes were identified during the peptide binding.

The preliminary results in this chapter show that the FBAR is a promising technology for the label-free measurement of conformational changes.

---

[6] Parts of this chapter are included in Nirschl, M; Vörös, J; Ottl, J; Conformational Changes of Calmodulin on Calcium and Peptide Binding monitored by Film Bulk Acoustic Resonators. Accepted for Publication in Biosensors **2011**.

8.2 Experimental Section

8.2.1 FBAR Read-out

The FBARs used in this study are described in Chapter 3.2 but with a difference in the read-out: In addition to the resonant frequency, the admittance at resonance was measured and used as a measure of the viscoelastic dissipation. We assume that the change in admittance represents mainly a change in the motional resistance, which is a measure of the energy dissipated by the adsorbate [21]. As a consequence, the measured change in admittance at the serial resonant frequency is referred to as motional resistance throughout this chapter.

A flow cell (about 60µl volume) was mounted on top of the FBAR; all measurements were carried out at a flow of 10 µl/min using a peristaltic pump. When changing liquids the flow was increased to 900 µl/min for 10 seconds to ensure a compete exchange of the liquids in the flow cell.

8.2.2 Reagents and Materials

Neutravidin was obtained from Pierce (Rockford, IL, USA). Biotinylated calmodulin and Ca^{2+}/calmodulin kinase II inhibitor (CaMKII 281–309) were purchased from Calbiochem (San Diego, CA, USA) and used within two weeks of thawing. Biotin, HEPES, NaCl, $CaCl_2$ and EDTA were all purchased from Sigma Aldrich (Germany). In all measurements 10 mM HEPES-NaOH, pH 7.2, 100 mM NaCl was used as buffer.

8.3 Results and Discussions

8.3.1 Immobilisation of Neutravidin and Biotinylated Calmodulin

Neutravidin (1 mg/ml) was adsorbed in flow directly on the cleaned gold surface. The binding of the Neutravidin to gold could be either attributed to physisorption or to sulphur-metal interaction [21, 194]. After a buffer rinse, biotinylated Calmodulin at concentrations of 30 nM, 300 nM and 3 µM was added.

Figure 49 shows the typical frequency and motional resistance behaviour during the Neutravidin adsorption and subsequent biotinylated Calmodulin binding. For both processes, a decrease in frequency, indicating the adsorption of mass to the surface, occurs concurrently with an increase in motional resistance, which reflects an increase in dissipation caused by the viscoelastic character of the adsorbed molecules. 20 µM biotin was then added to saturate the remaining Neutravidin binding sites.

Table 8 summarises the adsorbed bio-CaM at three concentrations. At the highest concentration (3µM) a surface mass of 190 ng/cm^2 bio-CaM (i.e. 11.4 pm/cm^2) was immobilised on the Neutravidin surface. This corresponds to 1.17 molecules of calmodulin per immobilised neutravidin.

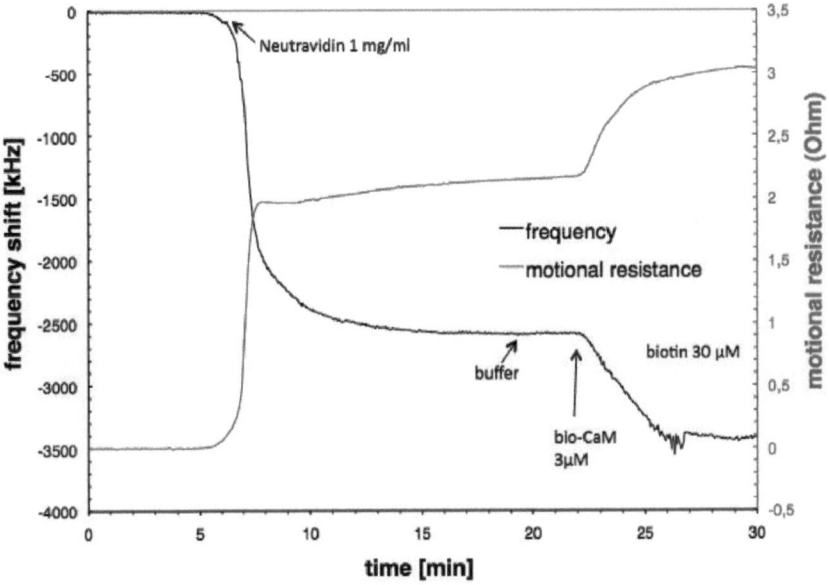

Figure 49: Resonant frequency shift and motional resistance change during Neutravidin adsorption and binding of biotinylated calmodulin.

8.3.2 Calcium Induced Conformational Changes of Calmodulin

Calcium chloride was injected in increasing concentrations over the calmodulin surface while monitoring the resonant frequency and the motional resistance. EDTA was used to remove the calcium from the calmodulin. Figure 50 shows the frequency and motional resistance change during calcium adsorption and desorption for the highest calmodulin concentration (300 μM). Upon calcium adsorption (blue arrows) both the frequency shift (black line) and the motional resistance (red line) decrease. With the addition of EDTA, the resonant frequency is restored to the original level and the motional resistance recovers to close to the original level, albeit with a slight drift downward. The behaviour of the motional resistance was similar (i.e. decrease with calcium adsorption) for all calmodulin concentrations. This decrease of the motional resistance is due to a decrease in acoustic energy dissipated in the calmodulin layer. From literature it is known that calmodulin transfers from a flexible to a more rigid molecule when calcium is bound and thus adsorbs less acoustic energy [195].

In a reference measurement, calcium was injected to a biotin saturated Neutravidin surface; the change in frequency and motional resistance was below noise (data not shown).

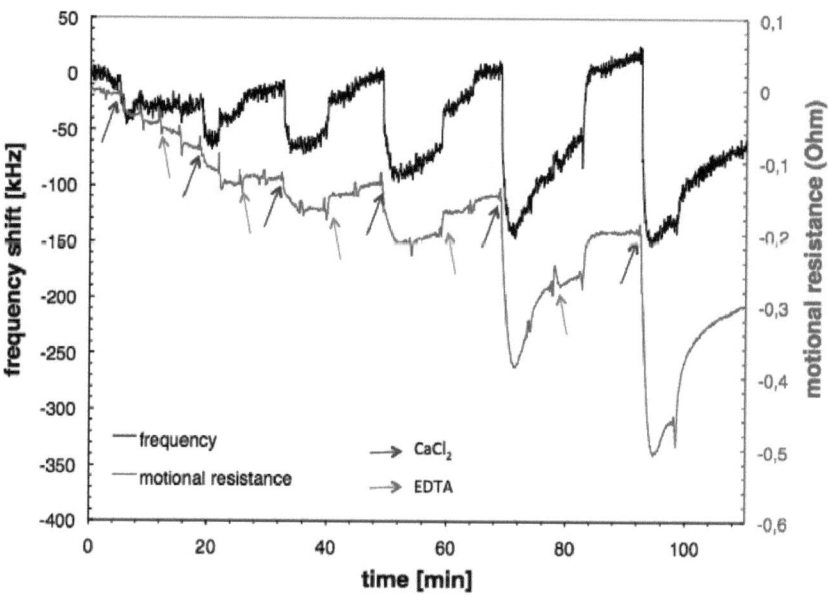

Figure 50: Frequency and motional resistance caused by the calcium adsorption. Blue arrows show the injections of $CaCl_2$, grey arrows the injections of EDTA injections

While the behaviour of the motional resistance was qualitatively equal for all calmodulin surface concentrations, the behaviour of the resonant frequency was found to be highly dependent on the surface concentration of calmodulin. In Figure 51 the ratio of the frequency shift caused by the adsorption of calcium at 1 mM to the frequency shift of the calmodulin adsorption is plotted against the surface density of the calmodulin (black lines and symbols). The equivalent surface mass, obtained by converting the resonant frequency shift of the calcium adsorption into mass per area assuming that the frequency shift is only influenced by mass changes on the surface is also shown (grey lines and symbols). For calmodulin densities lower than 150 ng/cm^2, the frequency shift is positive on both QCM and FBAR. Because the shift is positive, this cannot be explained with the mass of the bound calcium because an increase of mass causes a negative frequency shift on QCM and FBAR. The positive shift is likely to be the result of a conformational change as suggested previously [21, 22]. We suggest that a change in the tertiary structure in the calmodulin decreases the amount of water bound to the calmodulin and causes the increase in the resonant frequency.

For the highest surface density (190 ng/cm^2) the frequency shift is negative, indicating that more mass is bound to the sensor surface. This mass bound to the surface, however, cannot be only the mass of the adsorbed calcium, which is too small to significantly contribute to the shift. The mass effect of calcium bound to all 4 binding sites of all available calmodulin molecules would contribute only -8 kHz to the frequency shift, whilst the measured change is -168 kHz. We therefore suggest that both the negative and the positive frequency shift are caused by changes of the mass of water that moves together with the calmodulin. Coupled water within biomolecular layers is known to significantly contribute to the frequency shift on acoustic resonators [180, 196-198]. The coupled water also explains the behaviour of the frequency shift: At a lower protein surface density, the amount of water coupled inside the biomolecular layer is higher than at high surface concentrations. As an example, the ratio of the mass of the bound water to the mass of adsorbed streptavidin was found to vary from 7 to 1.5 with increasing surface density [199]. Similar conditions might exist for the calmodulin layer. Unfolding of the calmodulin molecule which exposes hydrophobic domains on calcium binding [200, 201] is likely to decrease the water content inside the layer, causing an increase of the resonant frequency. For higher calmodulin surface concentrations this effect decreases due the lower water content of the calmodulin layer. On the contrary, the elongation of the calmodulin upon calcium binding [202] increases the amount of water bound to the surface through two effects: Firstly, the increased thickness of the layer causes a larger water layer to be moved together with the acoustic resonator and secondly, the mass density of the layer is decreased by the elongation, and in this way more space is made available for water molecules. In addition, the increase of the thickness of the calmodulin layer alone will increase the wavelength of the acoustic wave in the resonator. The frequency shift of the elongation of a protein layer (typic from 5.8 nm to 6.2 nm was calculated to be -36 kHz using the model by Mason [124, 125]. This means that together with the -8 kHz shift from the calcium mass -124 kHz would result from a change in coupled water.

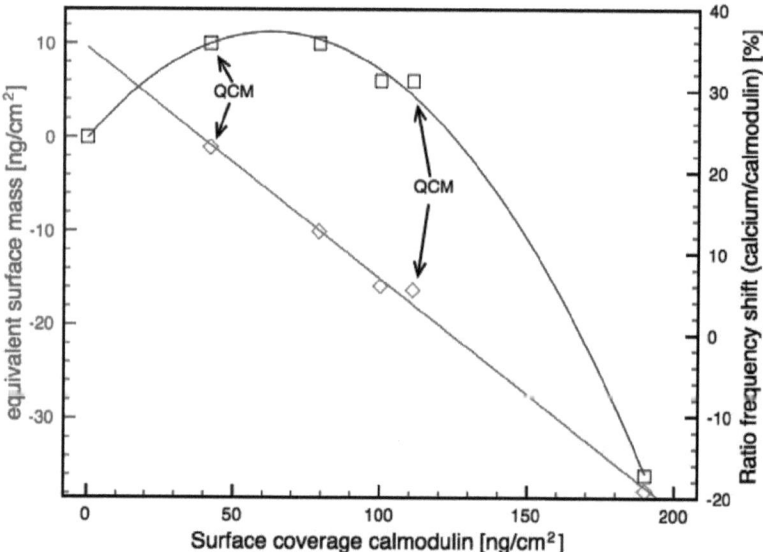

Figure 51: The frequency response caused by calcium adsorption for different surface densities of calmodulin. The ratio of the frequency shift of the calcium adsorption to the frequency shift of the calmodulin adsorption is plotted in black. The equivalent surface mass calculated from the frequency shift is plotted in red. The values for the QCM were taken from [22] and [21]. A line was fitted through the points showing the ratio of the frequency shifts and parabola through the points showing the equivalent surface mass.

At calmodulin concentrations of 300nM and 3µM, titration curves for calcium concentrations ranging from 2 µM to 1mM were measured. Figure 52 shoves the titration curve for a calmodulin concentration of 300 nM. The line shows the results of a fit of the data to a Langmuir isotherm to obtain the apparent dissociation constant K_d. The K_d obtained for the surface concentration of 300 nM was 14.27 µM, within the range from 3 to 22 µM for the four different calcium binging sites to reported in the literature [203]. The K_d using the highest surface concentration of the calmodulin was measured to be 296 µM and thus one order of magnitude higher than the values from literature. Apparently, the K_d is overestimated if the surface concentration is too high. One reason for this might be steric hindrance. For K_d the surface density of the calmodulin should not be too high but still in a range where the obtained frequency shifts are significant enough to be measured. Judging from Figure 51 a calmodulin surface density of around 70 ng/cm^2 would be expected to give reasonable K_d results together with a reasonable signal-to-noise ratio of the resonant frequency.

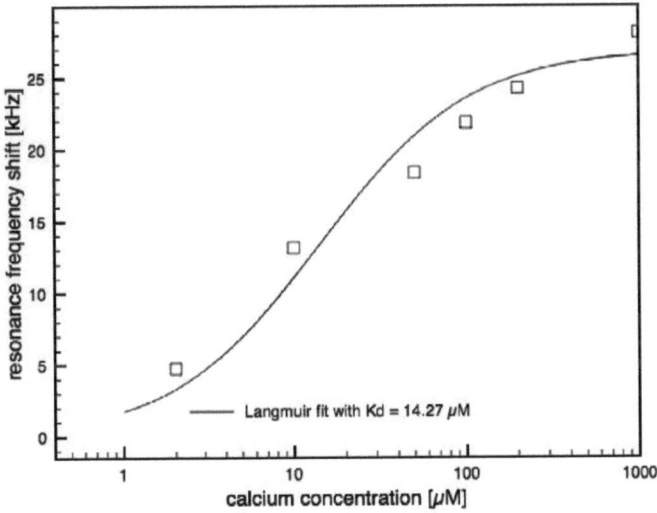

Figure 52: Frequency shift of calcium binding to calmodulin at different calcium concentrations. The line shows the results of a Langmuir isothermal plot with a fitted K$_d$ of 14.27 µM.

Measured on	Concentration bio-CaM [nM]	Bio-Cam bound to surface [ng/cm^2]	Binding efficiency of bio-CaM to Neutravidin	Mass change on addition of 1 mM CaCl$_2$ [ng/cm^2]	Ratio surface mass CaCl$_2$/bio CaM
FBAR	30	79	0.45	10	12.7 %
FBAR	300	100	0.69	6	6.0 %
FBAR	3000	190	1.17	-36	-19.1 %
QCM (Data from [22])	30	111	0.7	6	5.5 %
QCM (Data from [21])	3000	42	n.a.	10	23.1 %

Table 8: Summary of results for the bio-CaM binding and CaCl$_2$ adsorption of the FBAR measurements together with values from QCM literature ([21] and [22]).

8.3.3 CaMKII Peptide Binding to the Ca^{2+}/Calmodulin Complex

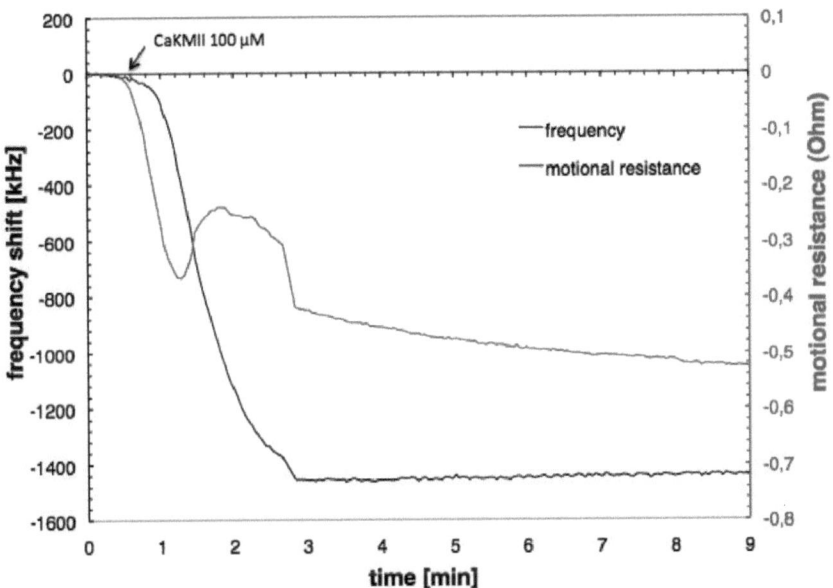

Figure 53: Frequency response and motional resistance change of CaMKII peptide binding to the Ca^{2+}/Calmodulin complex.

Figure 53 shows the binding of the peptide CaMKII to ApoCaM. The arrow shows the point at which the peptide was injected. Both the buffer and the CaMKII solution contained 1 mM CaCl$_2$.

The injection of the peptide is followed by a decrease of both frequency and motional resistance. The change of motional resistance starts promptly after the injection, whereas the frequency changes follow with a delay of about 45 seconds. This behaviour indicates that the mass adsorption is preceded by a conformational change of the calmodulin. While the resonant frequency shows normal binding behaviour, as seen for Neutravidin or calmodulin adsorption, the intial decrease of the motional resistance stops and reverses for about 90 seconds before again beginning a downward progression. A further conformational change causing changes in the viscoelastic properties of the adsorbed molecules might be the origin of this behaviour.

The resonant frequency shift at saturation (-1.4 MHz) is too large to be caused purely by the mass of the peptide. The frequency shift expected when one peptide binds to each calmodulin molecule on the surface is only -161 kHz. The additional contribution might be either caused by the peptide increasing the amount of water with in the calmodulin layer or by a large amount of water coupled to the peptide itself.

8.3.4 Conductance versus Frequency Shift Plots

In an attempt to show the adsorption processes in an easier and more comprehensible way, figures plotting motional resistance against frequency shift are used to describe the Neutravidin adsorption and subsequent biotinylated calmodulin binding, the binding of calcium to, and removal from, the calmodulin, and the binding of the peptide to the Ca^{2+}/Calmodulin complex.

Figure 54 a shows the adsorption of Neutravidin and the binding of biotinylated calmodulin to Neutravidin. Both adsorption processes are the binding of a soft material to the sensor surface. The mass of the soft biomolecules bound to the surface causes a frequency decrease, and due to the viscoelastic character of the molecules, they dissipate acoustic energy thus increasing the motional resistance. In the motional resistance versus frequency plots, the adsorption starts from the origin proceeds into the second quadrant.

Binding of calcium to calmodulin (for the highest calmodulin concentration) leads to a decrease in both frequency and resistance. The frequency decrease in this case is caused by additional water coupled to the resonator, and the decrease in motional resistance is caused by a rigidification of the calmodulin. In the motional resistance versus frequency plots (Figure 54 b) the adsorption starts at the origin and proceeds into the third quadrant. When calcium is bound to low surface densities of the calmodulin, the curve extends into the fourth quadrant (diagram not shown).

Figure 54 c shows the plot for the peptide binding. Adsorption starts with a decrease in frequency and resistance, as observed with calcium adsorption, indicating a conformational change. Then the curve turns towards an increase in motional resistance as found with the adsorption of Neutravidin (Figure 54 a). At the end of the adsorption, the curve again indicates decreasing motional resistance. From the form of the curve we suggest that the mass adsorption is both preceded and followed by a conformational change of the calmodulin.

a)

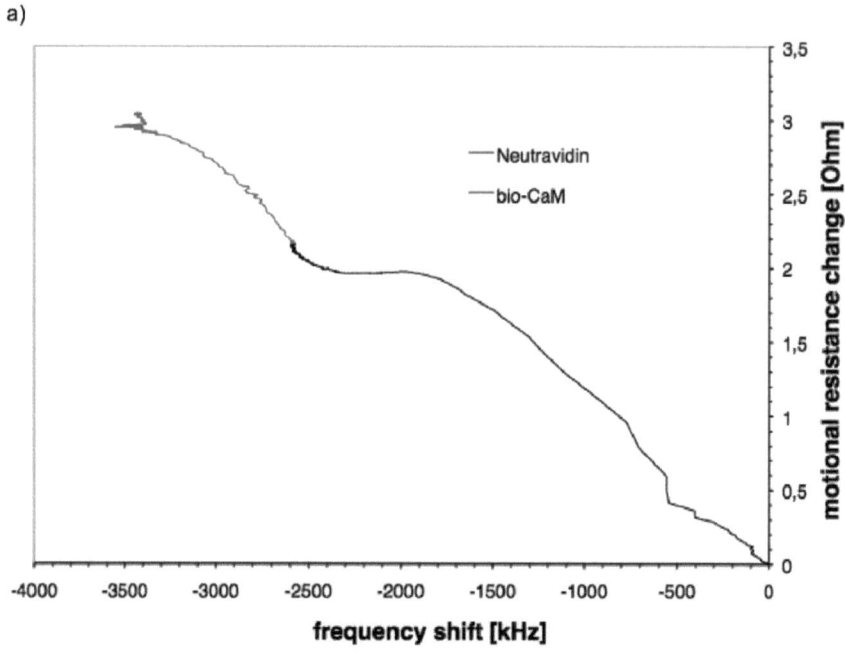

b)

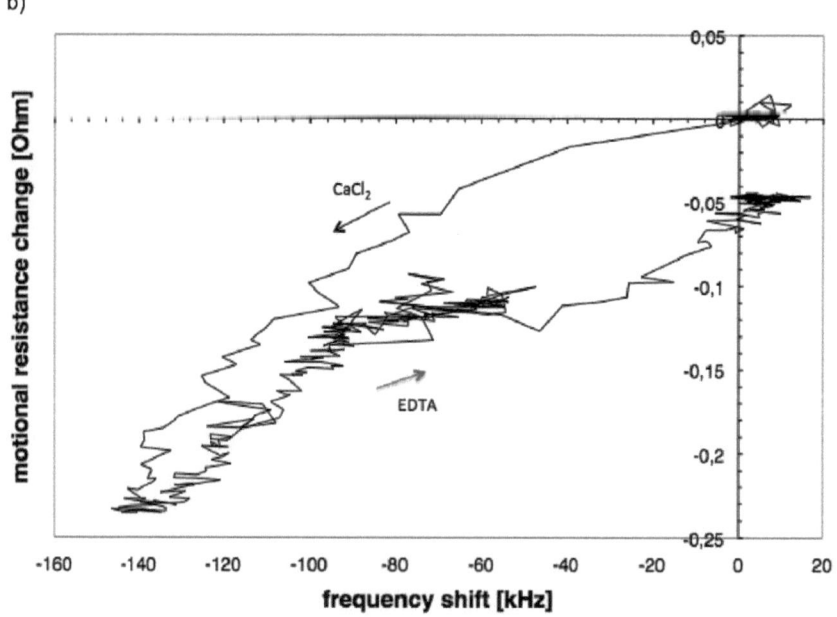

c)

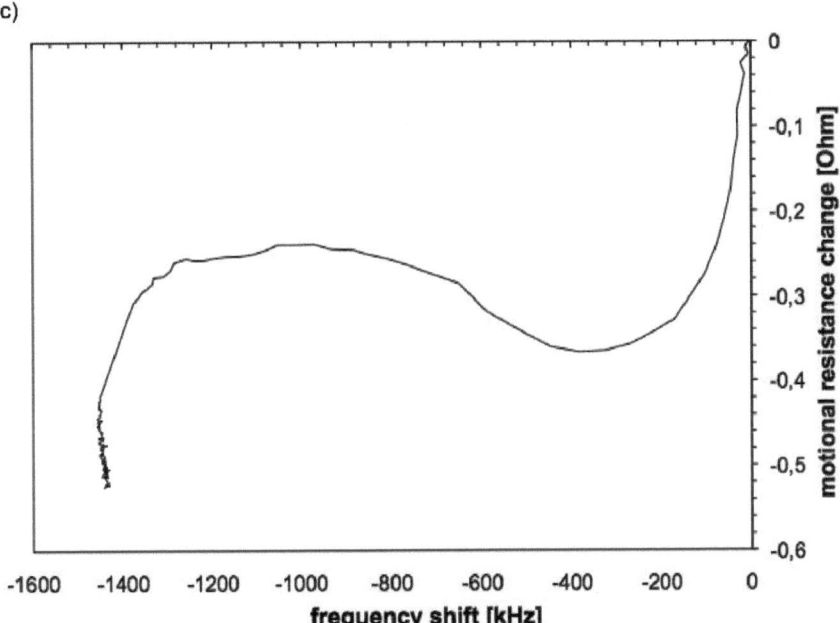

Figure 54: The adsorption of (a) Neutravidin and the binding of calmodulin, (b) the binding and removal of calcium to/from the calmodulin (c) and the peptide binding to the Ca^{2+}/Calmodulin plotted in resonant frequency versus motional resistance plots.

8.4 Conclusion

In this chapter the FBAR was used to monitor the conformational changes of calmodulin caused by calcium and peptide binding. Even though the limit of detection of this technology is lower than comparable technologies, and would be too low to detect the mass of the bound calcium, the signals caused by the conformational change were sufficient to determine an apparent K_D for the binding. The dependence of the signals on the surface concentration shows the importance of a careful assay development in order to obtain accurate K_D values.

In addition to calcium adsorption, conformational changes were also discovered during the binding of a peptide to the Ca^{2+}/Calmodulin complex. During these assays, three different processes could be obseved; the peptide binding, a mass adsorption and a conformational change. The different processes are more easily distinguished when plotting the motional resistance versus the resonant frequency shift and omitting the time axis.

Judging from these first measurements investigating conformational changes, the FBAR is a promising transducer for the investigation of conformational changes in a high throughput format with low sample volume. However a better understanding of the FBAR behaviour is necessary in order to utilize this technology in routine measurements, in particular it needs to be clarified how coupled water, stiffening or other phenomena influence the FBAR measurement results.

9 Conclusions and Outlook

In this thesis FBARs have been evaluated as label-free biosensors for use in biomolecular interaction analysis. To this end, a wide range of other label-free sensors such as optical and other acoustic sensors were presented in Chapter 1. Their key parameters such as the sensitivity, sample consumption and their ability to perform multiplexed sensing were summarised. As determined in comparison measurements of the FBAR with the SPR and the QCM in Chapter 5.2 and Chapter 7.2, the FBAR has a mass sensitivity, in terms of the limit of detection, comparable to the QCM but is at least one order of magnitude poorer than the most sensitive commercial transducers, the SPR and the DPI. However, in Chapter 4 it was shown that microstructured materials such as CNTs might significantly increase the device's mass sensitivity. However, even at current levels of sensitivity, the FBAR demonstrated a robust performance for protein and DNA detection (Chapter 5). Furthermore, the FBAR has additional strengths, such as its small size and the resulting high number of FBARs that can be integrated on a small chip. In Chapter 7 a FBAR chip containing 64 pixels was presented. In a first test measurement all available pixels were simultaneously read out during detection of DNA hybridization. The large number of pixels has the advantage that many different surface functionalisations can be used at the same time. This is vital if several substances need to be detected from a single sample, as in the case of diagnostic applications where one sample (e.g. a drop of blood) needs to be tested for a number of disease markers. While the maximum number of pixels presented in this thesis was 64, the FBAR technology is easily scalable, i.e. the number of pixels could be easily increased to some hundreds or thousands at reasonable costs.

Another advantage is the smaller liquid consumption per pixel. In the measurement in Chapter 7 the consumption per pixel was around 10 µl, and probably could be reduced to around 1 µl per pixel by reducing the size of the flow cell. An injection volume of around 50 µl seems to be achievable.

In Chapter 7.5 compartments were built around the individual FBAR pixels, further reducing the required liquid volme to the nl-range per pixel. This low volume might be of advantage if many tests need to be performed with a limited amount of sample liquid. Even though the test with compartments was only preliminary due to the limited number of available chips, measurement of the FBAR during spotting seems feasible. This would mean a significant increase of measurement speed which is an important feature in applications such as drug discovery and drug development where a high number of potential drug candidates need to be tested against a drug target in a short time.

In Chapter 8 the FBAR showed a remarkable performance while monitoring the conformational change of calmodulin during adsorbtion of calcium and binding of a peptide. The possibility of measuring the adsorption of calcium, despite the mass of the bound calcium being too small to give a measureable signal, increases the range of possible applications of the FBAR into the spectrum of detection of small molecules. In addition, information about the conformational state adds value to the measurement results. The comparison of the FBAR with other technologies that are capable of detecting conformational changes shows that with the low sample consumption and the possibility of parallel measurement the FBAR technology might be superior.

However, most results in this thesis like the calmodulin measurements are of a preliminary nature. The measurements show the potential of the FBAR technology but there are still many open questions. The FBAR gave different signals dependent on the surface concentration of the calmodulin. The water content was given as an explanation but it is still unclear to what extent the mass of the coupled water, the length of the calmodulin molecule, the stiffness of the molecule or other factors that have not yet been taken into account contribute to the resonant frequency shift. To improve the understanding, comparison measurements with other sensors like the SPR, the QCM or AFM should be done and the measurement results should be compared with a variety of other molecules that undergo conformational changes.

In summary, while the FBAR might perform poorly if considering only its mass sensitivity, which is one to two orders of magnitudes lower than the most sensitive transducers, the FBAR has been shown to have many strong features: The small size and the resulting low liquid consumption, the large number of resonators that can be used in parallel, the high measurement speed and the possibility of detecting conformational changes. With these features, it seems to be likely that there are applications for which the FBAR is superior to the other established technologies. One of these applications, demonstrated in this thesis, might be biomolecular interaction analysis.

References

1. Zimmermann, B., C. Hahnefeld, and F.W. Herberg, *Applications of biomolecular interaction analysis in drug development.* TARGETS, 2002. **1**(2): p. 66-73.
2. Hintersteiner, M., et al., *Confocal Nanoscanning, Bead Picking (CONA): PickoScreen Microscopes for Automated and Quantitative Screening of One-Bead One-Compound Libraries.* Journal of Combinatorial Chemistry, 2009. **11**(5): p. 886-894.
3. Länge, K., B. Rapp, and M. Rapp, *Surface acoustic wave biosensors: a review.* Analytical and Bioanalytical Chemistry, 2008. **391**(5): p. 1509-1519.
4. Milyutin, E. and P. Muralt, *Thin Film Bulk Acoustic Wave Resonators for Gravimetric Sensing.* Nanosystems Design and Technology, 2009: p. 103-116.
5. Vörös, J., et al., *Optical grating coupler biosensors.* Biomaterials, 2002. **23**(17): p. 3699-3710.
6. Homola, J., *Present and future of surface plasmon resonance biosensors.* Analytical and Bioanalytical Chemistry, 2003. **377**(3): p. 528-539.
7. Bally, M., et al., *Optical microarray biosensing techniques.* Surface and Interface Analysis, 2006. **38**(11): p. 1442-1458.
8. Grieshaber, D., et al., *Electrochemical biosensors-Sensor principles and architectures.* Sensors, 2008. **8**(3): p. 1440ñ168.
9. Stewart, M., et al., *Nanostructured plasmonic sensors.* Chem. Rev, 2008. **108**(2): p. 494-521.
10. Yan, R., D. Gargas, and P. Yang, *Nanowire photonics.* Nat Photon, 2009. **3**(10): p. 569-576.
11. Cunningham, B. and L. Laing, *Microplate-based, label-free detection of biomolecular interactions: applications in proteomics.* Expert Review of Proteomics, 2006. **3**(3): p. 271-281.
12. Qavi, A., et al., *Label-free technologies for quantitative multiparameter biological analysis.* Analytical and Bioanalytical Chemistry, 2009. **394**(1): p. 121-135.
13. Barbulovic-Nad, I., et al., *Bio-microarray fabrication techniquesóa review.* Critical Reviews in Biotechnology, 2006. **26**(4): p. 237-259.
14. Fang, Y., *Label-free cell-based assays with optical biosensors in drug discovery.* Assay and drug development technologies, 2006. **4**(5): p. 583-595.

15. Lucklum, R. and P. Hauptmann, *Acoustic microsensors—the challenge behind microgravimetry.* Analytical and Bioanalytical Chemistry, 2006. **384**(3): p. 667-682.
16. Sauerbrey, G., *Verwendung von Schwingquarzen zur Wägung dünner Schichten und zur Mikrowägung.* Zeitschrift für Physik A Hadrons and Nuclei, 1959. **155**(2): p. 206--222.
17. Kanazawa, K.K. and J.G. Gordon, *Frequency of a quartz microbalance in contact with liquid.* Analytical Chemistry, 1985. **57**(8): p. 1770-1771.
18. Voinova, M., et al., *Viscoelastic acoustic response of layered polymer films at fluid-solid interfaces: Continuum mechanics approach.* Physica Scripta, 1999. **59**: p. 391.
19. Voinova, M., M. Jonson, and B. Kasemo, *Dynamics of viscous amphiphilic films supported by elastic solid substrates.* Journal of Physics: Condensed Matter, 1997. **9**: p. 7799.
20. Cooper, M.A. and V.T. Singleton, *A survey of the 2001 to 2005 quartz crystal microbalance biosensor literature: applications of acoustic physics to the analysis of biomolecular interactions.* J Mol Recognit, 2007. **20**(3): p. 154--184.
21. Wang, X., et al., *Conformational chemistry of surface-attached calmodulin detected by acoustic shear wave propagation.* Molecular BioSystems, 2006. **2**(3-4): p. 184-192.
22. Furusawa, H., M. Komatsu, and Y. Okahata, *In Situ Monitoring of Conformational Changes of and Peptide Bindings to Calmodulin on a 27 MHz Quartz-Crystal Microbalance.* Anal. Chem, 2009. **81**(5): p. 1841-1847.
23. Rodahl, M., et al., *Simultaneous frequency and dissipation factor QCM measurements of biomolecular adsorption and cell adhesion.* Faraday Discussions, 1997. **107**: p. 229-246.
24. Johannsmann, D., et al., *Viscoelastic properties of thin films probed with a quartz-crystal resonator.* Physical Review B, 1992. **46**(Copyright (C) 2010 The American Physical Society): p. 7808.
25. Keller, C. and B. Kasemo, *Surface specific kinetics of lipid vesicle adsorption measured with a quartz crystal microbalance.* Biophysical journal, 1998. **75**(3): p. 1397-1402.
26. Höök, F., et al., *Variations in coupled water, viscoelastic properties, and film thickness of a Mefp-1 protein film during adsorption and cross-linking: a quartz crystal microbalance with dissipation monitoring, ellipsometry, and surface plasmon resonance study.* Anal. Chem, 2001. **73**(24): p. 5796-5804.

27. Johannsmann, D., *Viscoelastic, mechanical, and dielectric measurements on complex samples with the quartz crystal microbalance.* Physical Chemistry Chemical Physics, 2008. **10**(31): p. 4516-4534.
28. Weigel, R., et al., *Microwave acoustic materials, devices, and applications.* IEEE Transactions on Microwave Theory and Techniques, 2002. **50**(3): p. 738-749.
29. Gizeli, E., et al., *A Love plate biosensor utilising a polymer layer.* Sensors and Actuators B: Chemical, 1992. **6**(1-3): p. 131-137.
30. Kovacs, G., et al., *A Love wave sensor for (bio) chemical sensing in liquids.* Sensors and Actuators A: Physical, 1994. **43**(1-3): p. 38-43.
31. Länge, K., et al., *Packaging of Surface Acoustic Wave (SAW) based Biosensors: an Important Issue for Future Biomedical Applications.*
32. Drafts, B., *Acoustic Wave Technology Sensors.* Sensors Magazine, 2000. **17**(10).
33. Gronewold, T., *Surface acoustic wave sensors in the bioanalytical field: Recent trends and challenges.* Analytica Chimica Acta, 2007. **603**(2): p. 119-128.
34. Benes, E., et al., *Comparison between BAW and SAW sensor principles.* IEEE Transactions on Ultrasonics, Ferroelectrics and Frequency Control, 1998. **45**(5): p. 1314-1330.
35. Bjurstrom, J., et al. *3I-5 Design and Fabrication of Temperature Compensated Liquid FBAR Sensors.* in *Proc. IEEE Ultrasonics Symposium.* 2006.
36. Dickherber, A., C.D. Corso, and W. Hunt, *Lateral field excitation (LFE) of thickness shear mode (TSM) acoustic waves in thin film bulk acoustic resonators (FBAR) as a potential biosensor.* Conf Proc IEEE Eng Med Biol Soc, 2006. **1**: p. 4590--4593.
37. Link, M., et al., *Solidly mounted ZnO shear mode film bulk acoustic resonators for sensing applications in liquids.* IEEE Trans Ultrason Ferroelectr Freq Control, 2006. **53**(2): p. 492--496.
38. Weber, J., et al., *Shear mode FBARs as highly sensitive liquid biosensors.* Sensors and Actuators A: Physical, 2006. **128**(1): p. 84 - 88.
39. Lakin, K.M., *Thin film resonators and filters.* Ultrasonics Symposium, 1999. Proceedings. 1999 IEEE, 1999. **2**: p. 895-906 vol.2.
40. Carlotti, G., et al. *Surface acoustic waves in c-axis inclined ZnO films.* in *Proc. Ultrasonics Symposium IEEE 1990.* 1990.
41. Fardeheb-Mammeri, A., et al., *c-axis inclined AlN film growth in planar system for shear wave devices.* Diamond and Related Materials, 2008. **17**(7-10): p. 1770 - 1774.

42. Link, M., et al. *C-axis inclined ZnO films deposited by reactive sputtering using an additional blind for shear BAW devices.* in Proc. IEEE Ultrasonics Symposium. 2005.
43. Wang, J.S. and K.M. Lakin. *Sputtered C-Axis Inclined ZnO Films for Shear Wave Resonators.* in Proc. Ultrasonics Symposium. 1982.
44. Akiyama, M., et al., *Influence of metal electrodes on crystal orientation of aluminum nitride thin films.* Vacuum, 2004. **74**(3-4): p. 699 - 703.
45. Martin, F., et al. *Shear mode coupling and tilted grain growth of AlN thin films in BAW resonators.* in Proc. IEEE Ultrasonics Symposium. 2005.
46. Yanagitani, T., et al., *Characteristics of pure-shear mode BAW resonators consisting of (1120) textured ZnO films.* IEEE Trans Ultrason Ferroelectr Freq Control, 2007. **54**(8): p. 1680--1686.
47. Wingqvist, G., J. Bjurstrom, and I. Katardjiev. *Shear mode AlN thin film electroacoustic resonator for biosensor applications.* in Proc. IEEE Ultrasonics Symposium. 2005.
48. Nirschl, M., et al., *CMOS-Integrated Film Bulk Acoustic Resonators for Label-Free Biosensing.* Sensors, 2010. **10**(5): p. 4180-4193.
49. Link, M., *Study and realization of shear wave mode solidly mounted film bulk acoustic resonators (FBAR) made of caxis inclined zinc oxide (ZnO) thin films: application as gravimetric sensors in liquid environments.* 2006, Université Henri Poincaré, Nancy I.
50. Nirschl, M., et al., *Film bulk acoustic resonators for DNA and protein detection and investigation of in vitro bacterial S-layer formation.* Sensors and Actuators A: Physical, 2009. **156**(1): p. 180-184.
51. Hoa, X.D., A.G. Kirk, and M. Tabrizian, *Towards integrated and sensitive surface plasmon resonance biosensors: A review of recent progress.* Biosensors and Bioelectronics, 2007. **23**(2): p. 151-160.
52. Harrison, D. and H. Kjellberg, *Segmenting a market in the making: Industrial market segmentation as construction.* Industrial Marketing Management, 2009.
53. Siontorou, C. and F. Batzias, *Innovation in biotechnology: moving from academic research to product development-the case of biosensors.* Critical Reviews in Biotechnology, 2010(00): p. 1-20.

54. Bergstrom, J., S. Lofaas, and B. Johnsson, *Matrix coating for sensing surfaces capable of selective biomolecular interactions, to be used in biosensor systems.* 1995, Google Patents.
55. Homola, J., S. Yee, and G. Gauglitz, *Surface plasmon resonance sensors: review.* Sensors and Actuators B: Chemical, 1999. **54**(1-2): p. 3-15.
56. Löfås, S. and B. Johnsson, *A novel hydrogel matrix on gold surfaces in surface plasmon resonance sensors for fast and efficient covalent immobilization of ligands.* Journal of the Chemical Society, Chemical Communications, 1990. **1990**(21): p. 1526-1528.
57. Hearty, S., et al., *Surface plasmon resonance for vaccine design and efficacy studies: recent applications and future trends.* Expert Review of Vaccines, 2010. **9**(6): p. 645-664.
58. Stenberg, E., et al., *Quantitative determination of surface concentration of protein with surface plasmon resonance using radiolabeled proteins.* Journal of Colloid and Interface Science, 1991. **143**(2): p. 513-526.
59. Homola, J., *Surface Plasmon Resonance Sensors for Detection of Chemical and Biological Species.* Chemical Reviews, 2008. **108**(2): p. 462-493.
60. Nelson, B., et al., *Surface plasmon resonance imaging measurements of DNA and RNA hybridization adsorption onto DNA microarrays.* Anal. Chem, 2001. **73**(1): p. 1-7.
61. Shumaker-Parry, J.S. and C.T. Campbell, *Quantitative Methods for Spatially Resolved Adsorption/Desorption Measurements in Real Time by Surface Plasmon Resonance Microscopy.* Analytical Chemistry, 2004. **76**(4): p. 907-917.
62. Boozer, C., et al., *Looking towards label-free biomolecular interaction analysis in a high-throughput format: a review of new surface plasmon resonance technologies.* Current Opinion in Biotechnology, 2006. **17**(4): p. 400-405.
63. Shumaker-Parry, J., et al., *Microspotting Streptavidin and Double-Stranded DNA Arrays on Gold for High-Throughput Studies of Protein- DNA Interactions by Surface Plasmon Resonance Microscopy.* Anal. Chem, 2004. **76**(4): p. 918-929.
64. Bravman, T., et al., *Exploring one-shot kinetics and small molecule analysis using the ProteOn XPR36 array biosensor.* Analytical biochemistry, 2006. **358**(2): p. 281-288.

65. Chang-Yen, D., D. Myszka, and B. Gale, *A novel PDMS microfluidic spotter for fabrication of protein chips and microarrays.* Microelectromechanical Systems, Journal of, 2006. **15**(5): p. 1145-1151.
66. Steiner, G., *Surface plasmon resonance imaging.* Analytical and bioanalytical chemistry, 2004. **379**(3): p. 328-331.
67. Flournoy, P.A., R.W. McClure, and G. Wyntjes, *White-Light Interferometric Thickness Gauge.* Appl. Opt., 1972. **11**(9): p. 1907-1915.
68. Do, T., et al., *A rapid method for determining dynamic binding capacity of resins for the purification of proteins.* Protein Expression and Purification, 2008. **60**(2): p. 147-150.
69. Abdiche, Y., et al., *Determining kinetics and affinities of protein interactions using a parallel real-time label-free biosensor, the Octet.* Analytical biochemistry, 2008. **377**(2): p. 209-217.
70. Cunningham, B., et al., *Colorimetric resonant reflection as a direct biochemical assay technique.* Sensors and Actuators B: Chemical, 2002. **81**(2-3): p. 316-328.
71. Tiefenthaler, K. and W. Lukosz, *Sensitivity of grating couplers as integrated-optical chemical sensors.* Journal of the Optical Society of America B, 1989. **6**(2): p. 209-220.
72. Ramsden, J., *Review of new experimental techniques for investigating random sequential adsorption.* Journal of Statistical Physics, 1993. **73**(5): p. 853-877.
73. Brusatori, M., Y. Tie, and P. Van Tassel, *Protein adsorption kinetics under an applied electric field: An optical waveguide lightmode spectroscopy study.* Langmuir, 2003. **19**(12): p. 5089-5097.
74. Höök, F., et al., *A comparative study of protein adsorption on titanium oxide surfaces using in situ ellipsometry, optical waveguide lightmode spectroscopy, and quartz crystal microbalance/dissipation.* Colloids and Surfaces B: Biointerfaces, 2002. **24**(2): p. 155-170.
75. Cross, G., et al., *The metrics of surface adsorbed small molecules on the Young's fringe dual-slab waveguide interferometer.* Journal of Physics D: Applied Physics, 2004. **37**: p. 74.
76. Cross, G., et al., *A new quantitative optical biosensor for protein characterisation.* Biosensors and Bioelectronics, 2003. **19**(4): p. 383-390.
77. Rothen, A., *The ellipsometer, an apparatus to measure thicknesses of thin surface films.* Review of Scientific Instruments, 1945. **16**: p. 26.

78. Tompkins, H. and E. Irene, *Handbook of ellipsometry*. 2005: William Andrew.
79. Westphal, P. and A. Bornmann, *Biomolecular detection by surface plasmon enhanced ellipsometry.* Sensors and Actuators B: Chemical, 2002. **84**(2-3): p. 278-282.
80. Kurrat, R., et al., *Plasma protein adsorption on titanium: comparative in situ studies using optical waveguide lightmode spectroscopy and ellipsometry.* Colloids and Surfaces B: Biointerfaces, 1998. **11**(4): p. 187-201.
81. Freire, E., *Isothermal titration calorimetry.* Current Protocols in Cell Biology, 2004. **17**: p. 1-17.8.
82. Sethi, R., *Transducer aspects of biosensors.* Biosensors and Bioelectronics, 1994. **9**(3): p. 243-264.
83. Macdonald, J., *Impedence Spectroscopy--Emphasizing Solid Materials and Systems.* Wiley-Interscience, John Wiley and Sons, 1987: p. 1-346.
84. Schöning, M. and A. Poghossian, *Bio FEDs (field-effect devices): state-of-the-art and new directions.* Electroanalysis, 2006. **18**(19-20): p. 1893-1900.
85. Bearinger, J., et al., *Electrochemical optical waveguide lightmode spectroscopy (EC-OWLS): A pilot study using evanescent-field optical sensing under voltage control to monitor polycationic polymer adsorption onto indium tin oxide (ITO)-coated waveguide chips.* Biotechnology and bioengineering, 2003. **82**(4): p. 465-473.
86. Brusatori, M. and P. Van Tassel, *Biosensing under an applied voltage using optical waveguide lightmode spectroscopy.* Biosensors and Bioelectronics, 2003. **18**(10): p. 1269-1277.
87. Kang, X., G. Cheng, and S. Dong, *A novel electrochemical SPR biosensor.* Electrochemistry Communications, 2001. **3**(9): p. 489-493.
88. Lavers, C., et al., *Electrochemically-controlled waveguide-coupled surface plasmon sensing.* Journal of Electroanalytical Chemistry, 1995. **387**(1-2): p. 11-22.
89. Ying, P., et al., *Adsorption of human serum albumin onto gold: a combined electrochemical and ellipsometric study.* Journal of Colloid and Interface Science, 2004. **279**(1): p. 95-99.
90. Wang, Z., et al., *Immunosensor interface based on physical and chemical immunoglobulin G adsorption onto mixed self-assembled monolayers.* Bioelectrochemistry, 2006. **69**(2): p. 180-186.

91. Yu, Y. and G. Jin, *Influence of electrostatic interaction on fibrinogen adsorption on gold studied by imaging ellipsometry combined with electrochemical methods.* Journal of Colloid and Interface Science, 2005. **283**(2): p. 477-481.
92. Marx, K.A., *Quartz crystal microbalance: a useful tool for studying thin polymer films and complex biomolecular systems at the solution-surface interface.* Biomacromolecules, 2003. **4**(5): p. 1099--1120.
93. Dong, Y., *The frequency response of QCM in electrochemically characterizing the immobilization on gold electrode.* Sensors and Actuators B: Chemical, 2005. **108**(1-2): p. 622-626.
94. Qi, H., C. Wang, and N. Cheng, *Label-free electrochemical impedance spectroscopy biosensor for the determination of human immunoglobulin G.* Microchimica Acta: p. 1-6.
95. Feynman, R. *Plenty of Room at the Bottom.* 1959.
96. Dahlin, A., et al., *High-Resolution Microspectroscopy of Plasmonic Nanostructures for Miniaturized Biosensing.* Anal. Chem, 2009. **81**(16): p. 6572-6580.
97. Larsson, E., et al., *Sensing characteristics of NIR localized surface plasmon resonances in gold nanorings for application as ultrasensitive biosensors.* Nano Lett, 2007. **7**(5): p. 1256-1263.
98. Rindzevicius, T., et al., *Plasmonic sensing characteristics of single nanometric holes.* Nano Lett, 2005. **5**(11): p. 2335-2339.
99. MacKenzie, R., et al., *Nanowire Development and Characterization for Applications in Biosensing.* Nanosystems Design and Technology, 2009: p. 143-173.
100. Sannomiya, T., et al., *Biosensing by Densely Packed and Optically Coupled Plasmonic Particle Arrays.* Small, 2009. **5**(16): p. 1889-1896.
101. Willets, K. and R. Van Duyne, *Localized surface plasmon resonance spectroscopy and sensing.* 2007.
102. Ghosh, S.K., et al., *Solvent and Ligand Effects on the Localized Surface Plasmon Resonance (LSPR) of Gold Colloids.* The Journal of Physical Chemistry B, 2004. **108**(37): p. 13963-13971.
103. Hutter, E. and J. Fendler, *Exploitation of localized surface plasmon resonance.* Advanced Materials, 2004. **16**(19): p. 1685-1706.
104. Svedendahl, M., et al., *Refractometric Sensing Using Propagating versus Localized Surface Plasmons: A Direct Comparison.* Nano Lett, 2009. **9**(12): p. 4428-4433.

105. Karlsson, R., et al., *Biosensor analysis of drug-target interactions: direct and competitive binding assays for investigation of interactions between thrombin and thrombin inhibitors.* Analytical biochemistry, 2000. **278**(1): p. 1-13.

106. Koehne, J., et al., *Miniaturized multiplex label-free electronic chip for rapid nucleic acid analysis based on carbon nanotube nanoelectrode arrays.* Clinical chemistry, 2004. **50**(10): p. 1886.

107. Gao, Z., et al., *Silicon nanowire arrays for label-free detection of DNA.* Anal. Chem, 2007. **79**(9): p. 3291-3297.

108. Elfström, N., et al., *Surface charge sensitivity of silicon nanowires: Size dependence.* Nano Lett, 2007. **7**(9): p. 2608-2612.

109. Patolsky, F., et al., *Electrical detection of single viruses.* Proceedings of the National Academy of Sciences of the United States of America, 2004. **101**(39): p. 14017.

110. Zheng, G., et al., *Multiplexed electrical detection of cancer markers with nanowire sensor arrays.* Nature biotechnology, 2005. **23**(10): p. 1294-1301.

111. Stern, E., et al., *Importance of the Debye screening length on nanowire field effect transistor sensors.* Nano Lett, 2007. **7**(11): p. 3405-3409.

112. Wanekaya, A., et al., *Nanowire-based electrochemical biosensors.* Electroanalysis, 2006. **18**(6): p. 533-550.

113. Bornhop, D., et al., *Free-solution, label-free molecular interactions studied by back-scattering interferometry.* Science's STKE, 2007. **317**(5845): p. 1732.

114. Bruylants, G., J. Wouters, and C. Michaux, *Differential scanning calorimetry in life science: thermodynamics, stability, molecular recognition and application in drug design.* Current medicinal chemistry, 2005. **12**(17): p. 2011-2020.

115. Saboury, A., *A review on the ligand binding studies by isothermal titration calorimetry.* Journal of the Iranian Chemical Society, 2006. **3**(1): p. 1-21.

116. Brown, M., *Introduction to thermal analysis: techniques and applications.* 2001: Springer Netherlands.

117. Lafitte, D., et al., *Cation binding mode of fully oxidised calmodulin explained by the unfolding of the apostate.* Biochimica et Biophysica Acta (BBA)-Proteins & Proteomics, 2002. **1600**(1-2): p. 105-110.

118. Larsson, C., et al., *Gravimetric antigen detection utilizing antibody-modified lipid bilayers.* Analytical biochemistry, 2005. **345**(1): p. 72-80.

119. Karlsson, R., *Affinity analysis of non-steady-state data obtained under mass transport limited conditions using BIAcore technology.* Journal of Molecular Recognition, 1999. **12**(5): p. 285-292.
120. Nel, A., et al., *Toxic potential of materials at the nanolevel.* Science, 2006. **311**(5761): p. 622.
121. Tappura, K., I. Vikholm-Lundin, and W.M. Albers, *Lipoate-based imprinted self-assembled molecular thin films for biosensor applications.* Biosensors and Bioelectronics, 2007. **22**(6): p. 912-919.
122. Vikholm-Lundin, I., et al., *Hybridization of binary monolayers of single stranded oligonucleotides and short blocking molecules.* Surface Science, 2009. **603**(4): p. 620-624.
123. Link, M., et al., *c-axis inclined ZnO films for shear-wave transducers deposited by reactive sputtering using an additional blind.* Journal of Vacuum Science & Technology A: Vacuum, Surfaces, and Films, 2006. **24**(2): p. 218-222.
124. Mason, W.P., *Piezoelectric crystals and their application to ultrasonics.* 1950, London, New York, UK: D.Van Nostrand Co., McMillan and Co.
125. Rosenbaum, J.F., *Bulk Acoustic Wave Theory and Devices.* 1988: Artech House, Inc.
126. Lu, C. and A.W. Czanderna, *Applications of Piezoelectric Quartz Crystal Microbalances.* 1984: Elsevier, New York, NY, USA.
127. Nakamura, K., H. Kobayashi, and H. Kanbara. *Evaluation of acoustic properties of thin films using piezoelectric overtone thickness-mode resonators.* in *Ultrasonics Symposium, 2000 IEEE*. 2000.
128. Lucklum, et al., *Thin film shear modulus determination with quartz crystal resonators.* 1999. **11**(2).
129. Mansfeld, G., et al., *Acoustic and acoustoelectronic properties of carbon nanotube films.* Physics of the Solid State, 2002. **44**(4): p. 674-676.
130. Weber, J., *Investigation of the Physical Properties, Performance and Application of MEMS Sensors Based on Bulk Acoustic Waves excited in Piezoelectric Thin Film Devices.* 2007, Universität Augsburg.
131. Jost, O., et al., *Rate-Limiting Processes in the Formation of Single-Wall Carbon Nanotubes: Pointing the Way to the Nanotube Formation Mechanism.* The Journal of Physical Chemistry B, 2002. **106**(11): p. 2875-2883.

132. Rinzler, A.G., et al., *Large-scale purification of single-wall carbon nanotubes: process, product, and characterization.* Applied Physics A: Materials Science & Processing, 1998. **67**(1): p. 29-37.
133. Wu, Z., et al., *Transparent, Conductive Carbon Nanotube Films.* Science, 2004. **305**(5688): p. 1273-1276.
134. Wenseleers, W., et al., *Efficient Isolation and Solubilization of Pristine Single-Walled Nanotubes in Bile Salt Micelles.* Advanced Functional Materials, 2004. **14**(11): p. 1105-1112.
135. Lucklum, R., C. Behling, and P. Hauptmann, *Gravimetric and non-gravimetric chemical quartz crystal resonators.* Sensors and Actuators B: Chemical, 2000. **65**(1-3): p. 277-283.
136. Vig, J.R., *On acoustic sensor sensitivity.* Ultrasonics, Ferroelectrics and Frequency Control, IEEE Transactions on, 1991. **38**(3): p. 311.
137. Avrekh, M., O.R. Monteiro, and I.G. Brown, *Electrical resistivity of vacuum-arc-deposited platinum thin films.* Applied Surface Science, 2000. **158**(3-4): p. 217-222.
138. Koski, K., J. Hölsä, and P. Juliet, *Deposition of aluminium oxide thin films by reactive magnetron sputtering.* Surface and Coatings Technology, 1999. **116-119**: p. 716-720.
139. Tait, R.N., et al., *Density variation of tungsten films sputtered over topography.* Journal of Applied Physics, 1991. **70**(8): p. 4295-4300.
140. Pum, D. and U. Sleytr, *Large-scale reconstitution of crystalline bacterial surface layer proteins at the air-water interface and on lipid films.* Thin Solid Films, 1994. **244**(1-2): p. 882-886.
141. Gy^rvary, E., et al., *Self assembly and recrystallization of bacterial S layer proteins at silicon supports imaged in real time by atomic force microscopy.* Journal of microscopy, 2003. **212**(3): p. 300-306.
142. Douglas, K., N. Clark, and K. Rothschild, *Biomolecular/solid state nanoheterostructures.* Applied Physics Letters, 2009. **56**(7): p. 692-694.
143. Shenton, W., et al., *Synthesis of cadmium sulphide superlattices using self-assembled bacterial S-layers.* Nature, 1997. **389**(6651): p. 585-587.
144. Moore, J., et al., *Creation of nanometer-scale patterns with selected metal films.* Applied Physics Letters, 2009. **72**(15): p. 1840-1842.
145. Dieluweit, S., D. Pum, and U. Sleytr, *Formation of a gold superlattice on an S-layer with square lattice symmetry.* Supramolecular Science, 1998. **5**(1-2): p. 15-19.

146. Mertig, M., et al., *Fabrication of highly oriented nanocluster arrays by biomolecular templating.* The European Physical Journal D-Atomic, Molecular, Optical and Plasma Physics, 1999. **9**(1): p. 45-48.
147. Wahl, R., et al., *Electron-Beam Induced Formation of Highly Ordered Palladium and Platinum Nanoparticle Arrays on the S Layer of Bacillus sphaericusNCTC 9602.* Advanced Materials, 2001. **13**(10): p. 736-740.
148. Mertig, M., et al., *Formation and manipulation of regular metallic nanoparticle arrays on bacterial surface layers: an advanced TEM study.* The European Physical Journal D-Atomic, Molecular, Optical and Plasma Physics, 2001. **16**(1): p. 317-320.
149. Wahl, R., et al., *Multivariate statistical analysis of two-dimensional metal cluster arrays grown in vitro on a bacterial surface layer.* Chem. Mater, 2005. **17**(7): p. 1887-1894.
150. Aichmayer, B., et al., *Small Angle Scattering of S Layer Metallization.* Advanced Materials, 2006. **18**(7): p. 915-919.
151. Schreiter, M., et al., *Functionalized pyroelectric sensors for gas detection.* Sensors and Actuators B: Chemical, 2006. **119**(1): p. 255-261.
152. Hüttl, R., et al., *Calorimetric methods for catalytic investigations of novel catalysts based on metallized S-layer preparations.* Thermochimica Acta, 2006. **440**(1): p. 13-18.
153. Lucarelli, F., et al., *Carbon and gold electrodes as electrochemical transducers for DNA hybridisation sensors.* Biosensors and Bioelectronics, 2004. **19**(6): p. 515-530.
154. Lucarelli, F., et al., *Electrochemical and piezoelectric DNA biosensors for hybridisation detection.* Analytica Chimica Acta, 2008. **609**(2): p. 139-159.
155. Jönsson, U., et al., *Real-time biospecific interaction analysis using surface plasmon resonance and a sensor chip technology.* Biotechniques, 1991. **11**(5): p. 620.
156. Steel, A., et al., *Immobilization of nucleic acids at solid surfaces: effect of oligonucleotide length on layer assembly.* Biophysical journal, 2000. **79**(2): p. 975-981.
157. Peterson, A., L. Wolf, and R. Georgiadis, *Hybridization of mismatched or partially matched DNA at surfaces.* J. Am. Chem. Soc, 2002. **124**(49): p. 14601-14607.
158. Kimura-Suda, H., et al., *Base-dependent competitive adsorption of single-stranded DNA on gold.* J. Am. Chem. Soc, 2003. **125**(30): p. 9014-9015.
159. Boozer, C., S. Chen, and S. Jiang, *Controlling DNA orientation on mixed ssDNA/OEG SAMs.* Langmuir, 2006. **22**(10): p. 4694-4698.

160. Auer, S., et al., *Detection of DNA hybridization in serum matrix by surface plasmon resonance and film bulk acoustic resonators.* 2010: p. unpublished results.
161. Vikholm-Lundin, I. and R. Piskonen, *Binary monolayers of single-stranded oligonucleotides and blocking agent for hybridisation.* Sensors and Actuators B: Chemical, 2008. **134**(1): p. 189-192.
162. Vikholm-Lundin, I., et al., *A comparative evaluation of molecular recognition by monolayers composed of synthetic receptors or oriented antibodies.* Biosensors and Bioelectronics, 2008. **24**(4): p. 1036-1038.
163. Jin, W., et al., *A DNA sensor based on surface plasmon resonance for apoptosis-associated genes detection.* Biosensors and Bioelectronics, 2009. **24**(5): p. 1266-1269.
164. Gong, P., et al., *Hybridization behavior of mixed DNA/alkylthiol monolayers on gold: Characterization by surface plasmon resonance and 32P radiometric assay.* Anal. Chem, 2006. **78**(10): p. 3326-3334.
165. Su, X., Y. Wu, and W. Knoll, *Comparison of surface plasmon resonance spectroscopy and quartz crystal microbalance techniques for studying DNA assembly and hybridization.* Biosensors and Bioelectronics, 2005. **21**(5): p. 719-726.
166. Su, H., S. Chong, and M. Thompson, *Interfacial hybridization of RNA homopolymers studied by liquid phase acoustic network analysis.* Langmuir, 1996. **12**(9): p. 2247-2255.
167. Kurosawa, S., et al., *Quartz crystal microbalance immunosensors for environmental monitoring.* Biosens Bioelectron, 2006. **22**(4): p. 473--481.
168. Bjurström, J., G. Wingqvist, and I. Katardjiev, *Synthesis of textured thin piezoelectric AlN films with a nonzero c-axis mean tilt for the fabrication of shear mode resonators.* IEEE Trans Ultrason Ferroelectr Freq Control, 2006. **53**(11): p. 2095--2100.
169. Lakin, K.M., *Thin film resonator technology.* IEEE Trans Ultrason Ferroelectr Freq Control, 2005. **52**(5): p. 707--716.
170. Wingqvist, G., et al. *Shear mode AlN thin film electroacoustic resonator for biosensor applications.* in *Proc. IEEE Sensors.* 2005.
171. Lucklum, R., *Non-gravimetric contributions to QCR sensor response.* The Analyst, 2005. **130**(11): p. 1465-1473.

172. Voinova, M.V., M. Jonson, and B. Kasemo, *Missing mass effect in biosensor's QCM applications.* Biosens Bioelectron, 2002. **17**(10): p. 835--841.
173. Rodahl, M. and B. Kasemo, *On the measurement of thin liquid overlayers with the quartz-crystal microbalance.* Sensors and Actuators A: Physical, 1996. **54**(1-3): p. 448-456.
174. Francis, L., et al., *In situ evaluation of density, viscosity, and thickness of adsorbed soft layers by combined surface acoustic wave and surface plasmon resonance.* Anal. Chem, 2006. **78**(12): p. 4200-4209.
175. Wingqvist, G., et al., *On the applicability of high frequency acoustic shear mode biosensing in view of thickness limitations set by the film resonance.* Biosensors and Bioelectronics, 2009. **24**(11): p. 3387-3390.
176. Keller, C.A., et al., *Formation of supported membranes from vesicles.* Phys Rev Lett, 2000. **84**(23): p. 5443--5446.
177. Horvath, R., G. Fricsovszky, and E. Papp, *Application of the optical waveguide lightmode spectroscopy to monitor lipid bilayer phase transition.* Biosensors and Bioelectronics, 2003. **18**(4): p. 415-428.
178. YuM, L. and G.B. Sukhorukov, *Protein architecture: assembly of ordered films by means of alternated adsorption of oppositely charged macromolecules.* Membr Cell Biol, 1997. **11**(3): p. 277--303.
179. Picart, C., et al., *Molecular basis for the explanation of the exponential growth of polyelectrolyte multilayers.* Proceedings of the National Academy of Sciences of the United States of America, 2002. **99**(20): p. 12531.
180. Caruso, F., D. Furlong, and P. Kingshott, *Characterization of ferritin adsorption onto gold.* Journal of Colloid and Interface Science, 1997. **186**(1): p. 129-140.
181. Grieshaber, D., et al., *Swelling and contraction of ferrocyanide-containing polyelectrolyte multilayers upon application of an electric potential.* Langmuir, 2008. **24**(23): p. 13668--13676.
182. Wågberg, L., G. Pettersson, and S. Notley, *Adsorption of bilayers and multilayers of cationic and anionic co-polymers of acrylamide on silicon oxide.* Journal of Colloid and Interface Science, 2004. **274**(2): p. 480-488.
183. Boulmedais, F., et al., *Buildup of exponentially growing multilayer polypeptide films with internal secondary structure.* Langmuir, 2003. **19**(2): p. 440-445.

184. Notley, S., M. Eriksson, and L. WÂgberg, *Visco-elastic and adhesive properties of adsorbed polyelectrolyte multilayers determined in situ with QCM-D and AFM measurements.* Journal of Colloid and Interface Science, 2005. **292**(1): p. 29-37.
185. Eichelbaum, F., et al., *Interface circuits for quartz-crystal-microbalance sensors.* Review of Scientific Instruments, 1999. **70**(5): p. 2537-2545.
186. Norling, M., et al., *Oscillators Based on Monolithically Integrated AlN TFBARs.* Microwave Theory and Techniques, IEEE Transactions on, 2008. **56**(12): p. 3209-3216.
187. Augustyniak, M., et al. *An Integrated Gravimetric FBAR Circuit for Operation in Liquids Using a Flip-Chip Extended 0.13¬øm CMOS Technology.* in *Solid-State Circuits Conference, 2007. ISSCC 2007. Digest of Technical Papers. IEEE International.* 2007.
188. Schneider, T., et al. *Fast impedance analyzer interface with direct-sampling-technique for highly damped resonant gas sensors.* in *Sensors, 2005 IEEE.* 2005.
189. Keller, C.A. and B. Kasemo, *Surface Specific Kinetics of Lipid Vesicle Adsorption Measured with a Quartz Crystal Microbalance.* 1998.
190. Tukkiniemi, K., et al., *Fully integrated FBAR sensor matrix for mass detection.* Procedia Chemistry, 2009. **1**(1): p. 1051-1054.
191. Reimhult, E., et al., *Simultaneous Surface Plasmon Resonance and Quartz Crystal Microbalance with Dissipation Monitoring Measurements of Biomolecular Adsorption Events Involving Structural Transformations and Variations in Coupled Water.* Analytical Chemistry, 2004. **76**(24): p. 7211-7220.
192. Gabl, R., et al., *First results on label-free detection of DNA and protein molecules using a novel integrated sensor technology based on gravimetric detection principles.* Biosens Bioelectron, 2004. **19**(6): p. 615--620.
193. Larsson, C., M. Rodahl, and F. Hook, *Characterization of DNA Immobilization and Subsequent Hybridization on a 2D Arrangement of Streptavidin on a Biotin-Modified Lipid Bilayer Supported on SiO2.* Analytical Chemistry, 2003. **75**(19): p. 5080-5087.
194. Lyle, E., G. Hayward, and M. Thompson, *Acoustic coupling of transverse waves as a mechanism for the label-free detection of proteinñsmall molecule interactions.* The Analyst, 2002. **127**(12): p. 1596-1600.
195. Komeiji, Y., Y. Ueno, and M. Uebayasi, *Molecular dynamics simulations revealed Ca2+-dependent conformational change of calmodulin.* FEBS letters, 2002. **521**(1-3): p. 133-139.

196. Höök, F., et al., *Energy Dissipation Kinetics for Protein and Antibody- Antigen Adsorption under Shear Oscillation on a Quartz Crystal Microbalance.* Langmuir, 1998. **14**(4): p. 729-734.
197. Höök, F., et al., *A comparative study of protein adsorption on titanium oxide surfaces using in situ ellipsometry, optical waveguide lightmode spectroscopy, and quartz crystal microbalance/dissipation.* Colloids and Surfaces B: Biointerfaces, 2002. **24**(2): p. 155-170.
198. Vörös, J., *The density and refractive index of adsorbing protein layers.* Biophysical journal, 2004. **87**(1): p. 553-561.
199. Höök, F. and B. Kasemo, *The QCM-D technique for probing biomacromolecular recognition reactions.* Piezoelectric Sensors, 2007: p. 425-447.
200. Zhang, M. and T. Yuan, *Molecular mechanisms of calmodulin's functional versatility.* Biochemistry and Cell Biology, 1998. **76**(2-3): p. 313-323.
201. Tan, R.-Y., Y. Mabuchi, and Z. Grabarek, *Blocking the Ca-induced Conformational Transitions in Calmodulin with Disulfide Bonds.* Journal of Biological Chemistry, 1996. **271**(13): p. 7479-7483.
202. Seaton, B., et al., *Calcium-induced increase in the radius of gyration and maximum dimension of calmodulin measured by small-angle X-ray scattering.* Biochemistry, 1985. **24**(24): p. 6740-6743.
203. Crouch, T. and C. Klee, *Positive cooperative binding of calcium to bovine brain calmodulin.* Biochemistry, 1980. **19**(16): p. 3692-3698.

Patents

While working on this thesis at Siemens AG and ETH Zurich I submitted 10 invention disclosures. Some of them are listed here, some are still pending:

Huber, T.; Nirschl, M.; Pitzer, D.; Schreiter, M., (EN) Device and Method for Detecting a Substance using a Thin Film Resonator (FBAR) Having an Insulating Layer. WO Patent, WO/2010/046,212: 2010.

Nirschl, M.; Schreiter, M.; Wersing, W., (DE) Device for detecting substance of fluid, has piezo-acoustic thin film resonator, piezo-electric layer, and electrode layer that is arranged on piezo-electric layer. DE Patent ,DE102008049462A1

Nirschl, M.; Schreiter, M.; Schröter, C.; Sickert, D.; (DE) Humidity sensor has piezoacoustic resonator element which has two electrode layers and piezoceramic layer arranged between electrode layers, where sensor layer is provided with hygroscopic material for sorption of water. DE Patent, DE102007047155A1

Nirschl, M.; Schreiter, M.; Schröter, C.; Sickert, D.; (DE) Humidity sensor for detecting water content in e.g. organic solvent, has resonator element and sensor layer coupled together in manner such that resonant frequency of resonator element depends on amount of water absorbed by sensor layer. DE Patent, DE102007047153A1

Huber, T.; Nirschl, M.; Pitzer, D.; Schreiter, M., (DE) Vorrichtung und Verfahren zur Detektion einer Substanz mit Hilfe eines Dünnfilmresonators mit Isolationsschicht. De Patent, DE102008052437A1

Publications

Summary of publications with content partly or completely included in this thesis:

Nirschl, M.; Reuter, F., Vörös, J. Review of Transducer Principles for Label-free Biomolecular Interaction Analysis. *Biosensors* **2011**.

Nirschl, M; Vörös, J; Ottl, J; Conformational Changes of Calmodulin on Calcium and Peptide Binding monitored by Film Bulk Acoustic Resonators. Accepted for publications in Biosensors **2011**.

Nirschl, M.; Sickert, D.; Schreiter, M.; Vörös, J. Frequency response of thin-film bulk acoustic resonators to the deposition of tungsten, platinum, aluminium oxide and carbon nanotube thin-films. *Accepted for Publication in Micro- and Nanosystems* **2010**.

Nirschl, M.; Schreiter, M.; Vörös, J. Comparison of FBAR and QCM-D sensitivity dependence on adlayer thickness and viscosity. *Sensors and Actuators A: Physical* **2010**.

Nirschl, M.; Rantala, A.; Tukkiniemi, K.; Auer, S.; Hellgren, A.-C.; Pitzer, D.; Schreiter, M.; Vikholm-Lundin, I. CMOS-Integrated Film Bulk Acoustic Resonators for Label-Free Biosensing. *Sensors* **2010**, *10*, 4180-4193.

Kyprianou, D.; Guerreiro, A.; Nirschl, M.; Chianella, I.; Subrahmanyam, S.; Turner, A.; Piletsky, S. The application of polythiol molecules for protein immobilisation on sensor surfaces. *Biosensors and Bioelectronics* **2010**, *25*, 1049-1055.

Auer, S.; Nirschl, M.; Schreiter, M.; Vikholm-Lundin, I. Detection of DNA hybridization in serum matrix by surface plasmon resonance and film bulk acoustic resonators. *Analytical and Bioanalytical Chemistry* **2010**.

Tukkiniemi, K.; Rantala, A.; Nirschl, M.; Pitzer, D.; Huber, T.; Schreiter, M. Fully integrated FBAR sensor matrix for mass detection. *Procedia Chemistry* **2009**, *1*, 1051-1054.

Nirschl, M.; Blüher, A.; Erler, C.; Katzschner, B.; Vikholm-Lundin, I.; Auer, S.; Vörös, J.; Pompe, W.; Schreiter, M.; Mertig, M. Film bulk acoustic resonators for DNA and protein detection and investigation of in vitro bacterial S-layer formation. *Sensors and Actuators A: Physical* **2009**, *156*, 180-184.

Mertig, M.; Bluher, A.; Erler, C.; Katzschner, B.; Pompe, W.; Nirschl, M.; Schreiter, M. Investigation of in-vitro bacterial surface layer formation by FBARs. *2009 IEEE Sensors* **2009**, 1161-1164.

Posters

Nirschl, M.; Vikholm-Lundin, I.; Auer, S.; Tukkiniemi, K.; Rantala, A.; Pitzer, D.; Vörös, J.; Schreiter, M. 64 Acoustic resonators on one chip for label-free bionsensing. *Advances & Challenges in Label-Free Technologies for Drug Discovery*, **2009**, San Diego, California, USA

Nirschl, M.; Vikholm-Lundin I.; Auer, S.; Tukkiniemi, K.; Rantala, A.; Pitzer, D.; Vörös, J.; Schreiter M. 64 Acoustic resonators on one chip for label-free bionsensing. *2nd European Summer School in Nanomedicine*, **2009**, Lisbon, Portugal

Thin Film Bulk Acoustic Resonators for Label-Free Detection of Biomolecules
Nirschl, M.; Vikholm-Lundin, I.; Auer, S.; Pitzer, D.; Huber, T.; Vörös, J.; Schreiter, M. Thin Film Bulk Acoustic Resonators for Label-Free Detection of Biomolecules. *10th World Congress on Biosensors*, **2008**, Shanghai, China

Oral Presentations

Nirschl, M.; Blüher, A.; Erler, C.; Katzschner, B.; Pompe, W.; Vikholm-Lundin, I.; Auer, S.; Vörös, J.; Schreiter, M.; Mertig, M. Film Bulk Acoustic Resonators for DNA and Protein Detection and Investigation of Bacterial S-Layer Formation. *22nd EUROSENSORS*. **2008**, Dresden, Germany

Nirschl, M.; Vikholm-Lundin, I.; Auer, S.; Erdel, T.; Pitzer, D.; Huber, T.; Vörös, J.; Schreiter, M. Thin Film Bulk Acoustic Resonators for Label-Free Biomolecule Detection. *2nd Label-Free Protein Array Workshop*, **2008**, ENS Cachan, Paris-France

i want morebooks!

Buy your books fast and straightforward online - at one of world's fastest growing online book stores! Environmentally sound due to Print-on-Demand technologies.

Buy your books online at
www.get-morebooks.com

Kaufen Sie Ihre Bücher schnell und unkompliziert online – auf einer der am schnellsten wachsenden Buchhandelsplattformen weltweit! Dank Print-On-Demand umwelt- und ressourcenschonend produziert.

Bücher schneller online kaufen
www.morebooks.de

 VDM Verlagsservicegesellschaft mbH
Heinrich-Böcking-Str. 6-8   Telefon: +49 681 3720 174   info@vdm-vsg.de
D - 66121 Saarbrücken   Telefax: +49 681 3720 1749 www.vdm-vsg.de

Printed by Books on Demand GmbH, Norderstedt / Germany